Get Set for Computer Science

Get Set for Computer Science

Alistair D. N. Edwards

Edinburgh University Press

To Anne. 'I'm glad I spent it with you.'

© Alistair Edwards, 2006

Transferred to Digital Print 2011

Edinburgh University Press Ltd
22 George Square, Edinburgh

Typeset in Sabon
by Servis Filmsetting Ltd, Manchester
Printed and bound in Great Britain by
CPI Antony Rowe, Chippenham and Eastbourne

A CIP Record for this book is available from the British Library

ISBN 0 7486 2167 9 (paperback)

CONTENTS

Part II: Thinking of a Computer Science Degree

PREFACE

If you have bought this book (or had it bought for you) then there is a high probability that you are seriously considering undertaking a Computer Science degree at university. Presumably you have had some experience of using computers, perhaps even programming them and, having enjoyed that, you are thinking that you would like to learn more about them and that you might choose a career related to the development or use of computers. Some people get to that stage and start filling in their university application (UCAS) form with little further thought about what they are letting themselves in for. The aim of this book is to give you some more food for thought before you start committing yourself. You need to know what Computer Science is, how it is taught in universities, how best to decide which universities to apply to, and then what it will be like to study on your chosen course.

In a busy life, it may be hard to find the time to read a book such as this. Yet, if you are thinking of committing three or more years of your life to the intense study of Computer Science, perhaps it is a worthwhile investment to take this time to find out more about what you will be letting yourself in for. However, if you really do not have the time to read the whole book, you might use it as a reference to dip into in order to find the essential information you need. For instance, you might keep it at hand when you are reading university prospectuses and web pages.

It is not easy to define Computer Science. 'The science of computation' is an easy phrase, but what is behind it? In fact there are many different interpretations of Computer Science, and these are reflected in the different courses that are offered in our universities. Deciding to study Computer Science at university thus involves two steps: first ensuring that it is the subject that you want to study, and then ensuring that you apply to departments that teach the 'flavour' of Computer Science most appropriate to your own interests.

It is vital to make sure you know what Computer Science in order to ensure that your current fascination with computers will be sustained through three or four (or more!) years of intense, academic study. Part I of this book will help you with that, giving brief descriptions of most of the component topics of Computer Science. It is important to be clear, though, that no degree course will cover all of those topics; there are just too many of them. Any course will have been designed to cover a subset of the topics and that is where Part I can help you to find the right course for you. Part II will also help you to choose the course and departments and to make your applications to them.

Part III takes the next step. It assumes that all has gone well, that you have found a place on an undergraduate course, and that you now have to make the most of the opportunities you find there.

It might seem that the last thing a sixth-form student needs is another book to read. The intention has been to keep this one short, but nevertheless it may seem too much to take on. It should still be worthwhile to dip into the book to pick out the bits of information you need. For instance, if a course description or prospectus contains subjects or terms which you do not understand, you should find an explanation in this book, probably in Part I.

In writing this book, the intention has been to make it as accessible as possible to its intended readers. On the one hand, the nature of the topic means the book has to be technical, but it is meant to be at a level that is understandable to 'A' level students. The book has to be accurate and correct and not over-simplified. Support must be given for some assertions and arguments and the contribution of others must be acknowledged. Thus, the normal academic conventions for citing references to other works is used. A name and date (for example, Edwards 1995) refers to the list of references at the end of the book, where you can find details of the original source of information (a book, in this example).

ACKNOWLEDGEMENTS

This book is the result of countless interactions with different people who have informed and influenced me. However, I would particularly like to mention: Anne Edwards, Norman Edwards, Connie Cullen, Matthew Ford, Roberto Fraile, Bill Freeman, Steve Gilroy, Andrew Green, Chris Heald, Richard Ho, Ben Mitchell, Matthew Naylor, Matthew Patrick, Stephen Ratcliffe, Louis Rose, Gus Vigurs, Iain Wallace, Andrew Walton and Chris Wright. Thanks also to Richard Fuller for generating Figure 6.1.

In addition, I would probably never have graduated with my degrees without the assistance of my supervisors: Tim O'Shea, Mary-Jane Irwin, Alan Gibbons and the late Ron Footit.

LIST OF FIGURES

LIST OF TABLES

PART I
Computer Science

1 THE ELEMENTS OF COMPUTER SCIENCE

This book is intended to be read mainly by people in secondary schools and sixth form colleges who are thinking about studying Computer Science in a British university. The book is divided into three parts. Part I is a brief general introduction to the subject of Computer Science. This will give you some idea what the discipline of Computer Science is in general. Part II then puts that information in the context of a university course.[1] It suggests how you should go about deciding whether a degree programme in Computer Science is really for you – and, if it is, how best to choose the programmes for which you will apply. Part III gives you an idea of what it will be like to study the subject in a university.

The first question to be addressed in this part is: what is Computer Science? The answer is not a simple one, which is why a whole part of the book is devoted to answering it.

The next question: Is Computer Science really a science? Not all university programmes even refer to the subject by the same title, as you will see in Section 8.1 of this book: Computer Science, Studies, IT . . .? This is largely a reflection of the confusion over the discipline.

Look at the phrase 'Computer Science'. There is a debate as to whether the second word should be 'Science', but what is inescapable is the first word: the subject is inextricably tied to computers, to a technology. Other subjects, History, Mathematics, Sociology and so on, have existed for a very long time – even if we did not begin studying them formally until relatively recently – but until there were computers, there was no Computer Science.

3

There are three prevalent views as to the nature of the discipline:

- Computer Science is a branch of engineering
- Computer Science is Mathematics
- Computer Science is a science.

This spectrum of interpretations exists in most 'scientific' disciplines. For instance, Mechanical Engineering relies on the science of Physics, the laws of which are expressed in Mathematics. The difference is that if you are choosing a university programme, you will decide what it is you want to study: Engineering, Physics or Maths. Whichever discipline you choose, you will know the main subject that you will be studying, even if you expect some exposure to each of the other two: an engineer also knows about Physics and Maths; a mathematician knows a lot about Maths, and something about its application to mechanical problems.

The difference with Computer Science is that it is used as an umbrella term. So, one university might offer a programme with that title, but the contents will be taught more from an engineering perspective, while a programme with the same name in another university might be almost entirely mathematical.

In this part of the book, we describe elements that can be expected to be found in a Computer Science programme. Once you know something about the content of each of these topics, you will have a better idea as to what style of programme is on offer in different universities, and to which programmes you are more likely to be attracted.

It is not possible to give an account of the structure that is Computer Science, because it does not have a structure. What is possible, instead, is to list and describe a number of topics that are considered to be part of the discipline, some subset of which you can expect to be covered in a Computer Science degree programme. One place to look for such a list of topics is the *Subject Benchmark for Computing* produced for the Quality Assurance Agency for Higher Education (McGettrick,

Arnott et al. 2000).[2] This is one of a series of documents that have been produced to 'provide a means for the academic community to describe the nature and characteristics of programmes in a specific subject' (ibid., Executive Summary). The benchmark statement states (p. 4), 'It is difficult to define Computing with any degree of precision.' So this part of the book is based on a list of topics given in the benchmark, which is acknowledged not to be a definition, as such, of the field.

The descriptions in this section will give you some idea of the material that might be covered in a module on the given topic within a Computer Science degree programme. The descriptions here cannot be definitive, though. Different departments and lecturers will have different opinions as to what should be covered within a module; the module will have to fit within the overall programme and there is only a finite amount of material that can be covered in the time allotted. At the same time the descriptions herein are brief. You will also notice that there is overlap between the subjects. For instance modules on Computer Networks, E-commerce and Web-based Computing will all make reference to the Web.

The intention is that, when you have read this section, module descriptions in prospectuses, webpages and programme specifications should make more sense to you (and those should be your main source of information about the programmes for which you might apply). Of course, you may want to know more about particular topics, and there are introductory texts available. Suitable texts relating to each topic are listed at the end of each subsection. Many of them are texts that might be used as the basis of a university module on the topic. Among them are some books in the Pearson Education *Essence of Computing* series. The greatest virtue of these books is their brevity (up to 200 pages). In other words, they are useful if you would like to know a bit more about any of the featured topics, without going to a full university text.

You must be aware that no Computer Science university programme will include all of these topics; there is not time for all of them! The point is that when you look at the details of the different programmes you will be able to see which subset of the topics is covered, and you can then try to choose a programme with whatever you consider the best balance of topics for you (see also Section 8.5: Choosing a Department).

The one development that pervades all advances in Computer Science is the continuing advance in computing power. Essentially the power of a computer is the amount of data that it can process in a unit time. It is hard to quantify this fairly. Computer manufacturers try to use the processor clock rate, measured in MHz, but that is not entirely meaningful as it does not account for how much is achieved in each clock tick (cycle). Another measure is the number of arithmetic operations that can be performed in a unit time.

However it is measured, processing power has undoubtedly increased over the history of Computer Science. This was famously characterised by Gordon Moore (co-founder of Intel Corporation) who pointed out in 1965 that the power of computer chips was doubling approximately every eighteen months (Moore 1965), an observation that has become known as *Moore's Law*. Every so often someone will state that the limits of Moore's Law are being reached, but then new technologies are found and developed, so that no limit has yet been met and Moore's Law has continued to hold. This constant increase in processing power pervades many of the developments of computer technology and the way it is taught.

NOTES

1. As used in British higher education, the term 'course' is ambiguous. It may refer to an entire degree programme (three or four years of study) or one of the component modules (for example, a set of lectures on a particular topic, lasting one semester). 'Programme' and 'module' are less ambiguous and they are the preferred terms in official documents. Thus, hereafter in this book 'programme' and 'module' are generally used, to avoid ambiguity.

2. Notice that the subject benchmark authors opted not for the common term 'Computer Science', but rather 'Computing'.

2 SOFTWARE

2.1 PROGRAMMING

Programming is central to Computer Science. A program[1] is a set of instructions that a computer can carry out. Those instructions must be precise and unambiguous and therefore are written in programming languages, which look like a mixture of Mathematics and English. These are artificial languages because they have been invented by people, but this means they have been defined to be precise and unambiguous. Programming languages have been defined that attempt to combine some of the readability of English with the precision of Mathematics. For instance, the program written in the language *Pascal* in Figure 2.1 would invite the user to enter a temperature in Fahrenheit and would then write out its equivalent in Celsius.

```
program FahrenheitToCelsius(input, output);
  { Program to convert a Fahrenheit
    temperature to its equivalent in Celsius }
  var
    Fahrenheit : real;
  begin
    write('Enter Fahrenheit temperature:  ');
    readln(Fahrenheit);
    writeln((Fahrenheit - 32) * 5 / 9, ' Celsius')
  end.
```

Figure 2.1 A very simple computer program, written in the language Pascal.

Notice how Figure 2.1 combines elements of English (words like 'begin', 'Fahrenheit') with mathematical notations. It is not important to understand this program, but it may help to know that the star (*) is used as a multiplication symbol; the conventional cross symbol ($\times$) is not used in programming

7

languages because it might be confused with the letter x. The text in curly braces {} is a *comment*. This is purely intended to be read by people; it is not part of the program as such. This program is entirely unambiguous; if you type in any Fahrenheit temperature it will always give the Celsius equivalent.

There are some basic features of programming that make it work. There is *branching*, whereby a program will take one of two or more alternative actions, depending on the information currently available. For instance, a variation on the program in Figure 2.1 would allow the user to convert either from Fahrenheit to Celsius or the other way round. The other important feature is *repetition* or *looping*. A program can loop around applying the same or similar operation to a set of data, refining the result each time. Most programming languages provide more than one form of branch and loop, because different forms are easier for the programmer to use in different logical contexts.

Languages also provide *procedures* and *functions* which are essentially a way of *abstracting* parts of the program's operation. That is to say, the programmer identifies a particular operation within the program, separates it and gives it a name. Thereafter the programmer uses that name to invoke that operation. For instance, the conversion from Fahrenheit to Celsius in the program in Figure 2.1 might be abstracted into a procedure, *FtoC*, as in Figure 2.2. The equivalent programme, which uses ('calls') the procedure *FtoC*, is shown in Figure 2.3.

```
procedure FtoC(Fahrenheit : real;
               var Celsius : real);
   begin
     Celsius := (Fahrenheit - 32) * 5 / 9
   end;
```

Figure 2.2 A Pascal procedure which will convert any temperature in Fahrenheit into Celsius.

To separate different parts of the logic into procedures is a good idea in terms of keeping the structure of the program clear, easy to understand and easier for programmers to

```
program temperatures(input, output);
   var
      Fahrenheit, Celsius : integer;
   begin
      write('Enter Fahrenheit temperature: ');
      readln(Fahrenheit);
      FtoC(Fahrenheit Celsius);
      writeln(Celsius, ' Celsius')
   end.
```

Figure 2.3 An equivalent program to that in Figure 2.1, but this one uses the *FtoC* procedure from Figure 2.2.

understand and maintain. Furthermore, it means that it is possible to reuse code. For instance, if one was writing a program which deals with a number of temperatures, it may be a frequent requirement to convert them between the different scales, and having the procedure *FtoC* available will make that easy. Programming largely consists of building new components (procedures) and then putting them together into larger structures. These in turn can be built into larger constructs, and so on. It is only this kind of structuring that makes it at all possible to build software of great complexity and size. It is sometimes referred to as a 'divide and conquer' approach. That is to say that each component (procedure) is quite simple, but when you combine them together you can create something that is very complex – and powerful.

There are a number of common misapprehensions about programming. As discussed in Section 2.3: Compilers and Syntax-Directed Tools, a programming language has a strict set of *syntax* rules regarding the legal order of the symbols of a language. For instance, in the fragments above, notice that most of the lines have a semicolon at the end. If any one of those semicolons were missing, the program would be syntactically incorrect and impossible to run. The program will be submitted to another program called a **compiler**,[2] which attempts to convert the program from this human-readable form into an **object code** program which can run on the computer. In order to do that, the compiler must first check the program's conformance to those syntax rules. The compiler cannot generate object code if the rules have not been met.

In other words, a syntactically incorrect program makes no sense. For example, there is no reasonable mathematical interpretation of:

```
(Fahrenheit − 32) */ 5 9
```

This contains the same symbols as the original expression, but in an illegal order. In such a case, the compiler would generate one or more error messages to help the programmer locate the specific error, and would not go on to produce the object code. New programmers often find it difficult to write a program that conforms to the syntax rules. They may feel as if they are in conflict with the compiler, that they try their best to get the syntax right, but the compiler is out to get them, to find as many errors as it can. Thus, when they finally get a program to compile, they think that their troubles are over – but they are not. Much harder than getting the syntax right is making the program do what it is supposed to – in all circumstances.

Anyone who has used a computer has at some time encountered an error, a 'bug', when the program does not do what it is supposed to. This can occur in most programs. Commercial software is extensively exercised and tested, and yet bugs still occur. The elimination of bugs is a vital part of the programming process. So, that process consists of:

1. Design the program.
2. Convert that design into a program in the chosen language.
3. Correct any syntax errors in the program, with the help of the compiler.
4. Test the program with many sets of data, correcting any errors which become apparent.

New programmers often do not give each of these phases the right attention. They will often skip (1), preferring to start straight away on writing program code. As mentioned above, they may then see (3) as the most important step and, having achieved it, they will often not spend sufficient time on (4), so that their final program is left with many bugs.

It is not possible to test a program on every possible set of input data. Even for the simple program in Figure 2.1, it would not be possible to test on every possible input value because the input, in principle, could be *any* number.[3] The art of testing is in devising a representative sample set of data that will stretch the program to its limits. For this example, it would make sense to check that the program will work if the input is 0, if it is negative, and if it is very large or very small. (In this case, the program probably will not work if it is *very* large, because any attempt to multiply a very large number by five is likely to generate a number that is larger than the computer can represent and hence lead to an error.) A successful test is one that does reveal an error, because it gives the programmer the information needed to eliminate a bug. The more bugs that are thus revealed, the fewer bugs that are likely to persist in the released program.

There is another, more subtle potential bug in this program. What if the program's user were to enter a temperature of −500°F? This is a valid number, so the program would print the corresponding number, but the temperature is below absolute zero, the lowest possible temperature. Is that a bug? This illustrates another important point about programming (and associated topics, such as Software Engineering): that of *specification*. The person who has commissioned the writing of the program is the expert in what it is required to do and ideally should provide a specification which is complete and unambiguous. In the case of the temperatures program, the 'customer' should have said how the program should handle impossible temperatures (those below absolute zero). The problem is that the customer does not always understand what is technically feasible and how to precisely define what they require.

As discussed above, part of programming is getting the program syntactically correct – in a form that will be accepted by the compiler. However, the syntax rules alone do not force the programmer to write in such a way that the operation of the program is obvious to another programmer. This is important because the vast majority of programming that goes on in the real world is not where the programmer is given a blank sheet and asked to write a program to meet some specification. Instead it is more likely to be a matter of modifying existing programs, either to give them new functionality or to fix bugs which have been uncovered.

One programmer reading another's program should be given as much help as possible to understand how it works. In fact, it is a common experience for a programmer to come back to a program that they wrote themselves some time ago only to find that they cannot quite remember why they did it that way. As a simple example, the program in Figure 2.4 is syntactically correct and performs the same operation as Figure 2.1, but is it as easy to understand? All that has been changed is the choice of names and the formatting in lines.

```
program t (input,output);var x:
real;begin write
('Enter Fahrenheit temperature: ');
readln(x);writeln((x-32)*5 /
9,' Celsius')end.
```

Figure 2.4 The same program as in Figure 2.1, but with different names and formatting. This program would compile, and perform the same operation, but would be more difficult for a *person* to understand.

Notice also that the program in Figure 2.1 included a comment (the text in curly brackets {}). This is included entirely for the benefit of programmers reading the code; it is not part of the program as such, in that its inclusion makes no difference to the way that the program runs. Good use of comments can greatly assist in the understanding of programs. It has even been suggested that, given the choice, comment without code is more useful than code without comment. In the former case you can always write the code when necessary; in the latter case you may not know what it was for, what it did or how it did it, in which case it is best to throw it away and start again.

Comments represent the first level of *documentation* of a program. First-year undergraduates are often surprised to find that in programming assignments the documentation is worth more marks than getting the program to work. However, that reflects the reality that good documentation is often hard to achieve but more important in the long run.

There are a large number of programming languages in existence and some of them and the differences between them are discussed in Section 2.2: Comparative Programming Languages. There is a constant debate as to which is the 'right' language for students to learn first. The important point is that the choice of language is not really that important; it is much more important that students learn good *principles* that can be applied in any language. Indeed, it is likely that you will learn more than one programming language within a Computer Science programme. Once you have learned the principles (taught using whichever language is adopted for that purpose in your department), then you should be able to apply them easily when writing programs in another language.

Students starting an undergraduate programme in Computer Science will have different levels of previous experience. Those who have done Computer Studies 'A' level or Computing Highers will probably have already done some programming, while others will not have done. Those who have programmed before will have an advantage in that it will not all be new to them, but even they may find that what they are expected to do in programming at university is different from what they have done before. On the other hand, those who have not programmed before should not be intimidated: the lectures will be pitched in the knowledge that this is the first programming experience for many of the class.

Programming has been the first topic covered – and in some detail – because it is central to Computer Science. It is not only important in that Computer Science is nothing without programs and computers, but you may realise that programming is very much about problem-solving. That sort of problem-solving approach will be common across many of the topics in a Computer Science programme (it is also a useful skill to have nurtured in *any* career that you might go on to after university). Bearing this background in mind, it is now possible to look at the other topics which make up the subject matter.

Further reading

Friedman, D. P., Wand, M. and Haynes, C. T. (2001) *Essentials of Programming Languages*, Cambridge, MA: MIT Press.

2.2   COMPARATIVE PROGRAMMING LANGUAGES

There are a large number of programming languages in existence. Programmers tend to discuss the merits of different languages with the sort of fervour that is usually reserved for discussions of religion and politics. In any university module it is not possible to study more than one or two of them in any depth. It can therefore be useful to study the set of languages, making comparisons and highlighting differences. This can help to illuminate different important properties, and it helps the Computer Scientist to make reasoned decisions about the language to use for a given application. The Church Turing Thesis (see Section 6.4: Theoretical Computing) states that all programming languages are equivalent in terms of what programs written in them can and cannot compute. In that sense – theoretically – it makes no difference which programming language a programmer should choose to use, but there are other reasons why different languages have been developed. Overall, the aim is generally to aid the creation of programs more quickly and with fewer errors, and different styles of programming language lend themselves to solving different kinds of problems.

There are a number of different styles of programming language, including:

- imperative
- functional

- logic
- object oriented.[4]

We will discuss each of these briefly below.

Imperative languages

In an imperative language, the programmer states exactly how the program is to achieve its desired result. It is a common requirement that programs sort items into order (for example, to get the names in a telephone directory into alphabetical order). There are a number of different methods or **algorithms** to do this and, when using an imperative language, the programmer will decide which one to implement and translate that algorithm into the language.

The majority of programs are written in imperative languages (see the discussion below). Increasingly these languages are also object oriented. These languages are designed around the conventional design of a computer system, the von Neumann architecture, illustrated in Figure 3.1, below. Central to imperative programming languages is the *variable*. This effectively refers to a location in the memory in which values can be stored. For example, the program fragment in Figure 2.1 uses the variable *Fahrenheit*. The programmer has *declared* that this variable will hold the value of a real number (that is, a number with a fractional part) in the lines:

```
var
    Fahrenheit: real;
```

In so doing the programmer has effectively allocated one of the locations in the memory to store that value and labelled it with the name *Fahrenheit*. The statement

```
readln(Fahrenheit);
```

reads a value from an input device (for example, the keyboard) and passes that value to be stored in the memory location that has been allocated. Every manipulation of that variable implies that its value must be accessed from memory into the processor, changed and then moved back to memory.

The language must provide features which support the concepts discussed in Section 6.3: Data Structures and Algorithms. Thus, for instance, simple variables are not the only data representation, but there are also aggregations of variables in *arrays* and *records*. It has already been mentioned that the fundamental control structures in a computer are branching and looping, but most programming languages provide more than one form of branch and loop because different forms are easier for the programmer to use in different logical contexts. We have also seen in the section on Programming how procedures and functions can be used to give structure to programs. Most languages also allow the programmer to collect procedures and functions into larger-scale structures, or *modules*.

You will see below that the non-imperative languages are often used for specialised purposes. In other words, imperative languages are the most general-purpose, used for many different purposes. Examples of imperative languages are C, Basic, Pascal and Ada.

Functional languages

In Mathematics a function is a *mapping* from the members of one set, called the *domain* to another set, called the *range*. For instance, a function f might be defined:

$$f(x): x \in \Re, x \to x^2$$

This says that the domain of f is the real numbers (that is, any number, including fractions) and the mapping is from those numbers to their squares. The same function might be implemented in a functional programming language. It might be given a more meaningful name, such as *square*, but the program would be written such that

```
square(x)
```

effectively refers to the square of the value of x. Another function (*plusone(x)*) might return the value of $x + 1$. These can be combined, so that the value $(x + 1)^2$ would be returned by

```
square(plusone(x))
```

An extension of these definitions is that a function might be applied to itself. For instance, the functional programmer might implement a function that returns x^4 by using square twice, thus:

```
square(square(x))
```

This leads to the concept of *recursion*, which is the idea of applying a function to itself. It has been noted above that a fundamental feature of programming is repetition. In imperative languages this is normally achieved by way of *loop* constructs (for example, 'Do the following ten times' or 'Keep doing this until all the elements are in order'). For instance, suppose the multiplication button on your calculator was broken and you wanted to multiply two numbers, x and y. You might achieve this by using just the *plus* button and adding x to itself y times. If you were writing a multiplication program in an imperative language that did not have a multiplication operator, then you might write the equivalent program that uses only the add operation in a loop multiple times. In a functional language, you would use recursion instead. To do so, you have to note that

$$xy = x + x(y - 1)$$

and that if $y = 0$, then $xy = 0$. This means that in a (fictional[5]) functional language, you might define a *multiply* function as follows:[6]

```
multiply(x, y) is
   if y = 0 return 0
   otherwise return x + multiply(x, y − 1)
```

Notice that the definition of *multiply* uses *multiply*, this is recursion. This may seem odd at first – like a dog chasing its own tail – but you must also note that there is a situation in which the definition is not recursive, that is when y is zero. Also note that the recursive definition is approaching the non-recursive definition (that is, as long as y is positive, then $y − 1$ gets closer to zero every time the function is invoked recursively). This means that *multiply* will keep invoking itself until it reaches the point at which $y = 0$.

The first ever functional language was *Lisp* (Winston and Horn 1989). This is a language designed for the processing of *list* data structures. This is very appropriate because a list has a recursive structure. Other functional languages are *Scheme* (Abelson, Sussman et al. 1985), *ML* (Paulson 1996) and *Haskell* (Thompson 1996). Traditionally, functional languages have been used in artificial intelligence and other research applications; they are not commonly used in commercial programming.

Logic languages

Logic programming is sometimes also called *declarative* programming in contrast to imperative programming. As discussed above, in an imperative program the programmer states exactly the method to be applied to achieve the desired result; in a logic language the programmer states what is the result that he or she wants to achieve, and it is up to the language as to how it achieves it. Take the example of sorting, for instance. Suppose you have a list (or *array*), *l*, of eight elements:

$$l = [10, 7, 8, 1, 12, 3, 2, 9]$$

in other words $l_1 = 10$, $l_2 = 7$, $l_3 = 8$ and so on. You might effectively specify in a declarative language that *sort* implies that for all values of i in the range one to seven and j in the range two to eight, that if $i \leq j$ then $l_i \leq l_j$. It would be up to the programming language implementer (that is, the one who writes the **compiler** or **interpreter** for the language) as to how that would be achieved.

As the name implies, logic programming depends on a form of logic – in this case that which is known as *predicate calculus*. In this calculus the programmer can formulate *propositions*, where a proposition is a logical statement which may or may not be true. For example, it is possible to assert that 'Fido is a dog', using the relation 'isa'

```
isa(fido, dog)
```

Furthermore one might state that a dog is an animal:

```
isa(dog, animal)
```

Now it would be possible to ask the question, 'Is Fido an animal?'

```
?isa(animal, fido)
```

which in this system, would return the result *True*. This is a very simple example. In fact, in a logic language one can perform much more powerful logical operations. For example, one might assert that *all* dogs are animals, or that it is true that every dog is either male or female. Of course, in practice logic languages are used for rather more challenging problems. The programmer can build up an extensive database of facts and then pose questions to that database. In other words, in a complex database it is possible for a logic programming system to search the entire database and to resolve any query posed.

As this description implies, one application of logic languages is in the management of databases, specifically *relational databases*. Another important application area is in *expert systems*. An expert system attempts to emulate expert human behaviour in a particular area. One application is medical, with expert systems being developed to perform diagnoses, in the way that doctors do (or, rather, in the way that they might do). Expert knowledge is captured ('mined') and expressed as a set of rules that are fed into the database. For example, rules might capture the knowledge 'If the patient has a temperature *and* a runny nose, *then* they have a cold.' Queries can then be formulated which describe an individual's symptoms and the logic programming system will work out the implications of those symptoms in terms of the rules it has, and hence attempt to make a diagnosis.

As discussed in Section 2.3: Compilers and Syntax-Directed Tools, computer programs cannot easily handle natural languages, such as English. However, logic programming offers one approach to attempting to do this. The kinds of rules used in formal language theory, in what is known as *context free grammars*, are easily expressed in logic languages, but beyond that, they can also deal with some of the complex features of natural languages which are not context-free. The most widely used logic language is *Prolog*.

Object oriented languages

The concept of object orientation has been around for a long time, but it is only relatively recently that its potential for writing programs quickly has been realised and its use become widespread. The central concept (as implied by the name) is the *object*. An object encapsulates items of data *and* the operations that can be performed on them. An object is an *instance* of a *class*. A class describes a set of entities and an object is one of those things. One database might use the class *person* to represent the characteristics of people that are relevant to the purposes of the database (for example, name, date of birth, gender). Each record in that database will contain the information for one particular individual, an object which is an instance of *person*. An object does not only contain data, though, it also defines the operations (or *methods*) that can be carried out on it. The most fundamental operation is to create an instance (an object) of that class. In this case that would correspond to adding a new person to the database, filling in the individual's details of name, birthday, gender and so on. One set of operations will extract that information (for example, to return the date of birth for the object). Others may perform more complex operations within an object (for example, calculating the person's current age).

A number of features make this a powerful approach to programming. Firstly, it seems to be an approach that matches the way that people (programmers) think. Another important concept is that of *inheritance* between classes. There may be a specific application which needs to use a more specific class which shares (inherits) the features of an existing class but adds new features to it. For example, a company might have a database of its employees. Employees are people, so the person class described above might be used, but the company will want to store information that is particularly relevant to it (National Insurance number, salary, job title and so on) and to define new operations (promote, increase salary and so on). Hence, a new class (*employee*) will be created which inherits all of the features of *person*, but which adds to it the features and operations needed for the company database.

The existence of classes can save programmers a lot of work because they can re-use existing ones. The programmers creating the company database might get the *person* class 'off-the-shelf' from an existing library, thereby saving themselves from having to write code to deal with those aspects of people that are common to us all, while they concentrate on the particular features and operations that the company needs in its database. The only downside of this approach is that the libraries available are vast and complex, so it can be a difficult job for the programmer to find the class which is closest to their requirements and to keep track of where all the components of the program exist.

Object oriented languages are good for most applications. Examples of object oriented languages are C++ and *Java*. C++ is an interesting example because it is based on the existing language, C. In fact it is a strict superset of C, so that a program written in C can be compiled with a C++ **compiler**. This is probably one of the reasons behind the popularity of C++ because programmers already familiar with C did not have to learn a whole new language to switch to the object oriented style.

Java is also similar to C and C++, but it was designed with a very specific objective. It is a language intended to be used on the Web. That is to say that a suitably enabled web browser (for example, Firefox or Internet Explorer) can run programs written in Java on an individual's PC (or *client*). This means that a webpage can do much more than just displaying text and pictures and links between them (all that is essentially possible with a conventional, HTML page) instead, it can be interactive. With power comes responsibility, though. If I am going to run programs on my PC that I have picked up simply by loading a webpage – created by any stranger on the net – then I do not want that program to be able to do just what it likes on my PC. Thus, Java was designed to be deliberately restricted in what it can do. A Java program cannot take over control of my PC and start re-writing portions of my hard disc, for instance.

The above classifications of languages are not hard and fast. You may have noticed, for example, that the logic language proposition *isa*, above, resembled a function in a functional language. In fact, imperative languages *may* include functions, so that it is possible to write a program in an imperative language in a functional style. Imperative languages usually allow recursion so that the programmer has the choice of writing using loops or recursion. Also, a language which is object oriented can also be imperative – or functional. Nevertheless, these distinctions between different styles of language are useful in investigating and comparing them.

Further reading

Sebesta, R. W. (1999) *Concepts of Programming Languages*,
Reading, MA: Addison-Wesley.

2.3 COMPILERS AND SYNTAX-DIRECTED TOOLS

We have seen above that there are a lot of different programming languages of different styles. They are all designed
to facilitate the expression of solutions to problems by programmers; programs are intended to be understandable by
people. A program starts off as a simple text file
containing code, such as that in Figure 2.1. In order for the
computer to perform the actions specified in the program it has
to be first translated into a set of instructions in a binary
code that operates on the computer **hardware**. This translation
is itself achieved by a program, called a **compiler** or **interpreter**.
   A recipe provides a reasonable analogy. Most recipe books
are aimed at cooks with some previous experience of cooking.
They therefore contain instructions written at a high level.
Such a recipe might include, for instance, the text in Figure 2.5:

'Make a white sauce with the butter, milk and flour'

Figure 2.5 Sample instruction from a recipe book, written at a *high level*.

The text in Figure 2.5 is based on the assumption that the
reader knows how to combine and cook those ingredients to
make a sauce. Another book which is aimed at less experienced cooks (children learning cooking, for instance) might
not work at such a high level, instead describing how to make
the sauce in much more detail, as in Figure 2.6.

'Heat the butter gently and then add the flour a bit at a time, taking care to thoroughly stir the flour in as you add it . . .'

Figure 2.6 Fragment of an instruction from a recipe book written at a
low level but presenting the equivalent instruction to that in Figure 2.5.

The text in Figure 2.6 is a lower-level description. It is not the lowest possible level, though, because it assumes that the reader understands words such as 'heat', 'butter' and so on.

Computers operate at a very low level. They *understand* nothing. They are mechanical with a set of rules built in that effectively say 'When in a given situation (or state), if the current item of data from the input is X, then generate output Y and move to a given new state.' Such simple rules are sufficient to implement operations such as adding two numbers together – and a set of such simple operations is in fact sufficiently powerful to implement any computer program. (See also the description of the Turing Machine in Section 6.4: Theoretical Computing.)

The lowest level of language used by computers is *machine code*. That is a set of **binary** signals (zeros and ones) which embody the kind of instruction in the previous paragraph. A small step up from machine code is *assembly code*. This has identical structure to machine code, but uses names and labels that are a bit easier for a person to read (and write) than zeros and ones. For example, the assembly code in Figure 2.7 would add two numbers and store the result.

```
LOAD  B
LOAD  C
ADD
STORE A
```

Figure 2.7 A fragment of a program written in assembly code.

LOAD, ADD and STORE are instructions in the set that the computer can perform. *A*, *B* and *C* represent locations in the computer's memory. Those locations are represented as numbers or *addresses*. In fact, LOAD, ADD and STORE correspond to numbers too (for example, LOAD might be represented as 100, ADD by 101, STORE by 102). In other words, the little program above could have been written as a set of numbers, but they would be hard for a person to remember, recognise and to write without error. Hence, assembly code (or 'assembly language') is meant as an aid for people to write programs that can run almost directly on the computer. I say 'almost' directly because there is one step that has to be gone through where another computer program takes the textual, assembly code and converts it into binary machine code. This program is called an *assembler* and is the simplest form of programming language translator.

Assembly language exists as a means of making it easier for a person to write computer programs. In practice it is still not a very convenient way of doing this; you can see that with the code in Figure 2.7 it is quite laborious just to add two numbers. Most programming is done at a much higher level, as we have seen in the little example program in Figure 2.1. You will have noticed that the language that the fragment was written in looks like a mixture of English and Mathematics, which is relatively easy for a trained programmer to write and understand. In order to run on a computer, that program has to be translated to the lower-level type of language in Figure 2.7 – or in fact to one level lower, machine code. This is another example of translation from a high-level language to machine code and there are computer programs that will do this for you. These are compilers The essential difference is that each instruction of assembly language translates directly to one instruction of machine code, but each instruction (or *statement*) of a high-level language will translate into a number of machine code instructions.

Programming languages are said to be *artificial*. That is to say that they have been invented by people, unlike natural languages (such as English and French) which have evolved. The useful feature of an artificial language is that precise rules can be given as to what is a legal construction in the language and what that construct will mean when the program is run on a computer. It is the existence of these rules which means that it is possible to write programs, such as compilers, to manipulate them. There is a good body of knowledge regarding the mathematical basis of the rules behind languages, known as *formal language theory*. This might be studied in a module on Theoretical Computing, but is also likely to be covered in a module on compilers and syntax-directed tools.

A *grammar* is a set of rules defining what are legal instructions in a language. This is also referred to as the **syntax** of the language. Take the example of the statement in the program in Figure 2.1:

```
write('Enter Fahrenheit temperature:');
```

This causes the message *Enter Fahrenheit temperature:* to appear on the computer screen when the program is run. The syntax rules for the language say that the brackets after the word *write* must contain something, because it makes no sense to write nothing on the screen. The first task of a compiler, therefore, is to make sure that all of the program follows the syntax rules of the given language. Only if there are no syntax errors can it generate machine code, which is why such tools are described as being 'syntax directed'.

Very few Computer Science graduates will ever be expected to write a compiler, but it is nevertheless useful for several reasons to study how to build one within a module. The development of formal language theory means that this is an area of programming with a good theoretical base and many tools to assist the programmer. Secondly, the kind of process involved in translation is similar to that found in many other applications. There are many other applications that require this kind of syntax-driven programming.

Further reading

Hunter, R. (1999) *The Essence of Compilers*, London: Prentice Hall.

2.4   COMPUTER VISION AND IMAGE PROCESSING

You may at some time have played *Kim's Game*. That is where you are presented with a tray of objects to look at for a short time. Then the tray is taken away and some of the objects removed. When it is brought back, you are expected to say which items were taken away. You probably would describe this as a memory game, but, if you think about it, there is much more involved than memory. When you are presented with the tray you look at the items on it. You have to tell them apart and then you will give them a name ('hammer', 'watch', 'cup' or whatever). Then, when you see the altered tray you will probably try to recall the visual image, but also the names of the items. In playing Kim's Game, therefore, you will be exercising many of the faculties which researchers in computer vision would like to give to computers. This example also illustrates one of the motivations for studying computer vision: that if we can make a computer do it, we may learn about how people and animals process visual information.

Computer vision is concerned with both low and high levels of processing. At the low level there is some form of representation of a visual image containing certain features which are important. The lowest level representation is a sequence of numbers representing each of the **pixels** of a picture. The number will represent the colour and brightness of the picture at that point. One pixel does not give us much information about the picture, but if a computer program examines the information for a group of pixels, it can pick out features from the image. For instance, *edges* are very important in images. You need to pick out the edges of an object to see its shape and then to recognise what it is. Finding the edges might seem simple, but it is not. You might look for a region with one colour surrounded by a different colour. However, it is unlikely that two adjacent pixels in any picture have completely different colour. In a picture an object is likely to have shadows around the edges, making them quite fuzzy. There are further complications that can come into effect: An edge may not be continuous. If it suddenly appears to come to an end, is it because the object of which it is a part is overlaid with another item? Or, perhaps it marks a change in the shape, a straight edge blending into a curve.

Low-level processing will deal with properties such as edges, colour and texture. You can observe some of the tools and transformations that can be carried out at this level in image processing programs, such as Adobe *Photoshop*. Here you can take an image from a scanner or **digital** camera and apply transformations to it. You may *sharpen* the image, which means that the software will find the edges in the image and then alter the pixels around those edges to make the edges less fuzzy. Photoshop is controlled by a user. That is to say, you choose the transformations you wish to apply and then decide whether you like the result; the software does the low-level processing but you do the high-level processing. However, there is software that attempts to do the high-level processing too. It has to be able to recognise objects and interpret them. For example, whatever you think about traffic congestion charges, it would not be possible to apply them if it were not for image processing software. The software is fed images from television cameras. It then has to work out what is in the picture, where there is a car, where the number plate is on the car, and then to read the numbers and letters off it.

The video image from a television camera is, of course, a moving picture. In other words it is effectively a series of still frames, each of which is slightly different from the previous one. This is an additional challenge to image processing. It is another example of a form of computing that is only possible with the great power of modern processors. Recall that a pixel might be represented by a set of numbers, but a high-definition picture includes millions of pixels per square inch, and a moving image might consist of thirty-two images per second. This represents a vast amount of data and **brute force**, and processing power is not the only part of the answer. Image processing **algorithms** have been developed which are very successful at compressing the amount of storage that an image requires. If an image has a region of pixels all of the same value (that is, an area of the picture of the same, uniform colour), there is no need to store each one individually, but just store the value once, along with an indication of the bounds of the area with that value. Similarly, in a video image of an object moving against a static background, the number of pixels that change between two frames may be small (that is, the pixels representing most of the background will be unchanged). Therefore, in the second frame, it is only necessary to represent information about the pixels that have changed since the previous one – the portion of the picture that has moved.

Sometimes the vast amounts of data involved can be handled simply by taking a lot of time to do the associated processing. Such extended processing is not always allowable, though. In some applications it is necessary to process images very quickly – in **real time**. For instance, a mobile robot may be guided entirely by its vision, by processing the input from cameras mounted on it. If the system is going to recognise an object in front of the robot as an obstacle with which it is on a collision course, it must do this quickly enough that the robot will be able to take avoiding action before the collision occurs.

One way of speeding up computation is to use specialised **hardware**, designed to implement graphical algorithms. Another hardware-based approach is to use multiple processors in parallel. Image processing is well suited to this approach in that it is possible to effectively give different parts of the image to different processors to work on with a relatively low requirement for coordination between them.

As is often the case with the exciting topics in Computer Science, vision and image processing relies on some quite heavy-duty Mathematics, including geometry. Anyone contemplating any involvement in computer vision or image processing (such as taking an undergraduate module on the topic) should not do so just because they are interested in the product (pretty pictures) unless they are fairly sure that they can cope with the underlying Mathematics.

That computer vision and image processing will continue to grow in importance is in little doubt. As we continue to generate more images, we must find better ways to process them, to extract the valuable information. Some examples have been mentioned above, but if you think of anywhere that images are generated, there is scope for automatically processing them, including medical images (x-rays and scans of various kinds), satellite pictures and biometric identification (for example, using technology to verify the identity of an individual by their face alone). An example of the continuing burgeoning of image generation is that there are millions of security CCTV cameras in the UK, but a much smaller number of people employed to look at the screens displaying their output. Vision systems are being developed to attempt to process these images and automatically detect suspicious behaviour. At the same time, satellites such as the Hubble Telescope are generating images of space much more quickly than current technology on earth can process them. The green men may be waving at us but we have not had time to develop the pictures!

Further reading

Schalkoff, R. (1989) *Digital Image Processing and Computer Vision*, New York: John Wiley.

2.5   PROGRAMMING FUNDAMENTALS

One of the favourite topics for discussion in any Computer Science department is which programming language to teach to first-year students. Programmers always have a favourite language and they often seem to believe that a good programmer can be nurtured by early exposure to the 'right' language, and irretrievably scarred by experience of a 'bad' one.[7] However, in truth, a number of fundamental aspects of programming are true regardless of the language that is used to express them. This is analogous to natural, spoken languages, as studied by linguists. The expression of a meaning in English is different from French, but there are characteristics of the languages which are similar: they each have a vocabulary, grammar and meaning (or *semantics*). The most profound fundamental property of programming languages is that if it is possible to write a program in one language, then it is possible to write it in any language. This is the *Church-Turing Thesis* (see Theoretical Computing). The reason why there are, nevertheless, many different languages is that it is easier to write some kinds of program in one (type of) language than it is in another.

Programming languages are invented and therefore somewhat easier to study than natural languages. For instance, the grammar (or **syntax**) of a language specifies legal utterances in the language. For example, the syntax rules of the programming language C would say that the following is a legal mathematical expression:

$$x = 3 + 4;$$

but that the following is *not* legal:

$$= x \ 3 + 4;$$

This *looks* wrong to a programmer or mathematician but also its meaning is not apparent. That is to say that its semantics are not defined. One advantage of inventing a language (such as a programming language) over natural languages (such as English) is that the language can be designed with precise rules. The rules of syntax are relatively easy to specify and can be done in such a way that software can be written to check that the syntax of any submitted program is legal (as discussed in Section 2.3: Compilers and Syntax-Directed Tools).

Semantic rules are harder to capture; there is a temptation to suggest that a program 'means what it means'. In other words, to find out what a program 'means' you run it on the computer and find out what happens. Programmers learn about the effects that different programming language constructs have and therefore how to make a program behave in the way intended, but sometime more formal, precise descriptions of a program's semantics are required, as discussed in Section 6.4: Theoretical Computing.

Programming is central to Computer Science. One way of teaching about programming is to study factors that can be said to be fundamental, true regardless of the programming language or languages used. Clearly, though, there is a lot of overlap between the subjects that would be covered in such a module and other topics listed in this section. For more details you might refer to the sections on Programming, Comparative Programming Languages, and Data Structures and Algorithms.

Further reading

Horowitz, E. (1987) *Fundamentals of Programming Languages*, Maryland: Computer Science Press.

2.6 SOFTWARE ENGINEERING

One of the historically most influential series of books on Computer Science is entitled *The Art of Computer Programming* (Knuth 1973). As the title implies, programming has often been seen as an art or craft, carried out by skilled artisans. However, when one looks at the products of programming – software that may be responsible for controlling an aeroplane or for processing millions of pounds – then an artistic, aesthetic approach is really not appropriate. An artist might be able to design a beautiful-looking bridge, but we would get an engineer in to make sure that the design is practical and safe. The same is increasingly true in the production of software. An engineering approach needs to be applied so that we can build software with reliable, predictable properties.

The problem is (as discussed above when questioning whether Computer Science is a science) that we do not have the same fundamental principles that (say) a construction engineer has. In designing a bridge, the engineer will fall back on the fundamentals of Physics, laws of forces, gravity, measurable properties of materials and so on, to create a design that can be tested in the abstract long before any concrete is poured or steel forged.

Computer software can be very complex, but then so is a jet engine. Both should ideally be designed and built in such a way that their behaviour is entirely predictable, but in practice software rarely is – as anyone who has had their PC crash will know. Bringing a certain amount of engineering-inspired discipline to the software production process is part of the approach. Different phases in the production process can be identified and tackled one by one. The most common approach is known as the *waterfall model*, as illustrated in Figure 2.8.

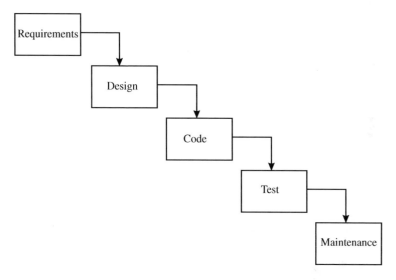

Figure 2.8 The waterfall model of software development. Each phase is clearly identified and completed before the next phase is undertaken.

Using a framework such as the waterfall model means that the different phases of development can be undertaken in a disciplined way. Thus, for instance, in the *Requirements* phase, the customer's exact needs will be identified. It is then up to the designers to generate a design that will fulfil all those requirements. Their design will then be coded in the chosen programming language by programmers. Testing should ensure that the code does implement the original requirements exactly. Then the product can be released to the customer, but there will still be a need to maintain it. That is to say that no matter how carefully the previous steps have been carried out, there will probably still be some bugs to deal with. Also, customers' requirements will change over time and part of the maintenance phase is to introduce modifications to the software to accommodate these changes.

As hinted above, the construction engineer will use Mathematics to prove the properties of a planned construction (that it can bear the loads that will be generated on it). Increasingly, mathematical approaches are also being used in Software Engineering. The design of a piece of program can be constructed so that it can be proved to be correct mathematically (using some of the techniques which might be studied under the heading of Theoretical Computing). It would seem ideal if every piece of software was thus designed and tested before being committed to computer code. Unfortunately, though, this is not practical because the effort to construct the mathematical description of a program and then to do the mathematical proof can be enormous; it can be a greater effort than writing the program itself. This technique currently tends to be used only for small pieces of critical code (for instance safety-critical code within a control program for an aircraft or nuclear generator).

Tools can be used to help in this process, such as theorem provers. These are not the only tools available to software engineers. Just as modern mechanical engineers use computer-aided design and manufacturing (CAD-CAM) tools, there are computer-aided software engineering (CASE) tools for the software engineer.

Further reading

Sommerville, I. (2004) *Software Engineering*, New York: Addison-Wesley.

NOTES

1. It is generally accepted that when referring to computer programs, the American spelling is used, even in the UK.
2. Words printed in bold are explained in the glossary.

3. This is not strictly speaking true. Computers are finite machines, where there is an infinite number of numbers, so they cannot all be represented. For most purposes, programmers cope with the fact that the computer can only represent a finite subset of all the numbers.

4. *Object oriented* is the accepted term for this style of programming, even though English purists might suggest that *object orientated* or even *object-orientated* might be more correct.

5. The use of fictional languages is quite common. They are referred to as *pseudo-code*, which is not another real language but a mixture of English and programming language syntax. Pseudo-code is generally used to illustrate points in a way that is independent of the language in which they are expressed.

6. Note that this particular definition will only work for positive values of *y*, but it could easily be extended to work for negative numbers too.

7. Witness the quote from the celebrated Computer Scientist, Edsger Dijkstra, who said, 'It is practically impossible to teach good programming to students that have had a prior exposure to BASIC: as potential programmers they are mentally mutilated beyond hope of regeneration.' (Dijkstra, E. W. (1982) *Selected Writings on Computing: A Personal Perspective*, Berlin: Springer-Verlag).

3 HARDWARE

3.1 COMPUTER ARCHITECTURE

If a builder is concerned with the bricks and components of a building, the architect is concerned with the structures into which the bricks can be constructed. So it is with computers: some people work at a low level, with the electronic components (as described in some of the other sections), whilst computer architecture is concerned with the ways in which they can be constructed into higher-level structures. The building architect may be concerned with the aesthetics of the building design, but the building should also fulfil its practical requirements (to be an efficient place in which to work, or live, or whatever). The computer architect will also be mainly concerned with the efficiency of the design, but might well argue that a good design is also elegant and pleasing.

Moore's Law has remained true in recent years, partly because of advances in the fabrication of electronic components. If elementary components are made smaller, they can be packed closer together, and therefore the distance that the electronic signal has to travel is shorter. However, it is necessary to ensure that the system can exploit the resultant speed, and that is largely up to the architectural design. That is to say, it is inefficient to have a component that can process n items of data per millisecond if other parts of the system can only deliver $n/2$ in that time.

There are some essential, high-level components of any computer:

- the processor (sometimes referred to as the CPU or *central processing unit*)

36

- memory
- communications.

Architecture is concerned with the design of these components and their interconnection. It also concerns the internal design of the processor. The processor consists of:

- a control unit
- registers, in which data can be stored
- an arithmetic and logic unit (ALU) in which data is combined and manipulated.

This architecture is illustrated in Figure 3.1. It is often referred to as the *von Neumann architecture*, because it is the architecture or design of one of the first computers, designed by John von Neumann, a design that has stood the test of time and is still the basis of most modern computers.[1]

A critical issue with computer system design is balancing the performance of all the components. Put simply, a computer might degrade to the speed of its slowest component unless the faster ones are put to good use, and not kept idle waiting for results from slower parts of the system. Memory is the simplest example. There are several different types of memory but in general the fastest memory is also the most expensive. It is thus not practical to have a system with vast amounts of high-speed memory. Instead, the system will have large amounts of cheap memory, and decreasing amounts of increasingly expensive but faster memory.

The fastest memory is in the form of *registers* on the processor itself. These can be accessed very quickly, but are inevitably limited in number. Then there is main memory, usually referred to as DRAM (for *dynamic random-access memory*). Finally, mass memory is usually provided on discs. This is very cheap, but very slow (physical movements of the disc take a long time compared to electronic speeds). There are gaps between each of these levels of memory which may be filled by intermediate forms. For instance, there may be small amounts of high-speed memory between the processor

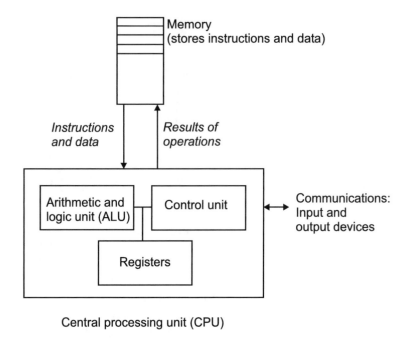

Central processing unit (CPU)

Figure 3.1 The architecture found on most computers, often referred to as the von Neumann architecture.

and DRAM. This is known as *cache*. The idea of cache is to have a small amount of fast memory available for frequently used data. An analogy would be the way most people deal with their food supplies. We do not go to the supermarket every time we need some food. Instead, we have a cupboard which we fill from the supermarket. Then, as we make our meals we access this convenient local supply (or cache) and only replenish it from the supermarket when we need to. We obtain a large amount of food on the supermarket shop, it takes a lot of time (compared to going to the cupboard), but we do it infrequently.

Knowledge of computer architecture is important to many areas of Computer Science. An example is in assessing the power of different processors. As mentioned earlier, anyone contemplating the purchase of a PC will have see advertisements which attempt to compare the power of different processors in megahertz.[2] This is not very meaningful. The frequency refers to the processor's *clock speed*. The processor includes a timing component known as its clock. If one operation on the processor can occur within a clock cycle, then a faster clock speed (that is, a larger number of megahertz) would imply a faster processor. However, it is not as simple as that. It may be that not every operation can be completed within one cycle. It is necessary to know a lot more about a processor's architecture to know the extent to which the clock speed is a reliable indication of its speed.[3]

Another important topic is the representation of data. Data is information reduced to a numerical form, and computers use numbers in **binary** form. This is because it is convenient to represent two distinct values. These are usually referred to as 0 and 1 but might be 'off' and 'on', or 'no' and 'yes'. Within an electronic computer, these values might be represented as 0 volts and 5 volts.

It is tempting to think that modern computers are so powerful that they can represent any information as accurately as we might ever need. The truth is that any representation is finite. No matter how many digits we store for a number, there is always scope for greater accuracy. For instance, if I want to store numbers of the orders of millions (decimal), then I need to use numbers with six digits. If I need larger numbers, say billions, then I need to use ten digits. The larger the numbers, the more digits are needed. There will always be a limit to the size that a given architecture can represent using this so-called *fixed point* representation. You will have noticed that pocket calculators use so-called 'scientific' notation, so that 10 billion might be represented as 1.0E10 (meaning 1.0×10^{10}). Using this kind of representation, it is possible to represent very large numbers – and very small ones (for example, the radius of an electron is 2.818E-15m). This representation is known as *floating point*. It is more complex than fixed point, because there are different components to the numbers. For instance, to calculate 1.23E11 $\times$ 2.998E8 will involve a multiplication (1.23×2.998) and an addition ($11 + 8$).

Floating-point numbers can be used to provide an alternative measure of processor speed: the *flop*, or floating-point operations per second. This is a measure chosen to be more realistic – and perhaps pessimistic – in that it is chosen precisely because a floating-point operation will usually require multiple processor cycles to complete, to manipulate the different components of the number. At the time of writing, the best supercomputers can achieve *teraflops*, or trillions of flops, that is, 1 teraflop = 10^{12} flop (you can find out the current 500 fastest supercomputers at http://www.top500.org/). Again, it is only with a real understanding of the architecture of processors that you will be able to make reasoned judgements about their relative strengths measured using such units.

There are two major types of processor architecture, known as *complex instruction set computers* (CISC) and *reduced instruction set computers* or RISC. In a CISC processor each instruction may perform a lot of work. That means that one instruction in a programming language might be translated into a small number of CISC instructions (see section 2.3: Compilers and Syntax-Directed Tools) while a RISC architecture would require a larger number of instructions to be executed to achieve the same result. The motivation behind the development of RISC architectures is the idea that it is easier to ensure that each of the (simple) RISC instructions can be made to execute very efficiently (quickly). It turns out that a large number of very efficient RISC instructions will generally execute more quickly than a small number of CISC instructions. In recent times much of the improvement in processor performance has been gained by making the best use of RISC architectures.

Further reading

Stallings, W. (2003) *Computer Organization and Architecture*, Upper Saddle River, NJ: Prentice Hall.

3.2   COMPUTER HARDWARE ENGINEERING

Most Computer Science programmes tend to concentrate on software topics, those related to programming, but software must have **hardware** on which to run. It is impossible to write programs without some understanding of how hardware works. Some programmes will give a superficial coverage of hardware, but others will give a much more in-depth level of teaching on hardware.

At the lowest level of electronics there are currents moving in wires and devices. The signals at this level are said to be 'analogue', or continuously varying. Some university modules will study the electronics at this level, though it is usually thought to be more the realm of the Electronic Engineer than the Computer Scientist. At this level there are components such as resistors and transistors. A current can be applied to the circuit and then voltages measured around different parts of the circuit. The nature of an analogue circuit is that a continuous change in one part of the circuit (such as gradually increasing a voltage from 0 to 3 volts) will cause corresponding continuous changes elsewhere. This behaviour can be predicted from circuit diagrams by applying simple mathematical rules – or the circuit can be built from physical components and direct measurements made.

Analogue electronics are used in applications such as musical performance. The singer produces an analogue voice signal (sound wave), that is picked up by a microphone which translates it into an electrical voltage. That voltage is amplified by an amplifier and fed into a speaker which converts the electrical signal back into a (powerful) sound signal. The analogue nature of this kind of interaction is apparent in that as the singer sings more loudly, so the amplified voice heard out of the speakers becomes louder.

As mentioned in several places in this book computers are based upon **digital** technology, not analogue iconology. The basis of digital technology is the simple **binary** number system. Binary numbers are composed of digits with one of two values: 0 or 1. Electronic circuits are used to represent these numbers. The components from which these circuits are made are referred to as *gates* (the name reflecting something that can be in one of two states, similar to a gate in a fence which can be either open or closed). All binary operations can be built from a small set of components, namely NOT, AND and OR gates. It is possible to draw diagrams of circuits consisting of these gates and to work out how they would behave – or one can physically connect real hardware components and observe how they behave. Increasingly, though, modules on hardware engineering will involve a third, intermediate possibility, that of *simulating* the circuit. That is to say, you draw a circuit diagram in a computer package and then you apply different inputs (0s and 1s) to the circuit and see what results are obtained on the output. This is a simple example of Electronic Computer-Aided Design, or *ECAD*.

Such gates and circuits are known as *combinatorial* because the output at any time depends on the input at that time. It is also necessary to have *sequential* elements, the output of which depends on the history of its inputs. In other words, sequential elements (often referred to as *flip-flops*) have *memory*. In practice, sequential circuits are built from combinatorial components.

One might question what these digital components have to do with the analogue electronics mentioned earlier. The connection is that digital circuits are built from analogue components. It is all a matter of interpretation. In a digital circuit it is only necessary to distinguish two values. Thus a voltage level of 0.0 volts may be defined to represent the binary value 0, while 5.0 volts is treated as representing 1. Using such *discrete* values explains much of the robustness of digital representations. In practice, it is very difficult to represent analogue voltages exactly. As a signal passes through electronic components it will be altered, so that a signal that is 5.0 volts in one part of the circuit may be reduced through *losses* in the circuit to 4.8 volts elsewhere. In an analogue representation 4.8 represents something different from 5.0. However, in a digital circuit any value from 0.0 to 2.4 volts might be treated as binary 0, while 2.5 volts to 5.0 would be treated as 1.

This treatment gives digital signals greater reliability since (inevitable) distortions in circuits can be ironed out.

All digital circuits are constructed from these simple components, but of course a computer component such as a processor is very complex. In other words, there may be millions of gates making up such a device. Such designs can only be achieved with the assistance of tools including ECAD software. Another form of support is provided by *Hardware Description Languages*. These allow the user to build a representation of a circuit design in a textual form, similar to a programming language. The importance of such descriptions is that they can be run through software tools that will verify them. In other words, long before the design is committed to hardware, the description can be checked to ensure that the circuit is correct, that it will have the expected behaviour. The cost of setting up tools to create a new computer chip is vast.[4] Hence, it is important to be as sure as possible that a circuit design is correct before it is implemented in hardware.

So, in any Computer Science programme, you will learn something about hardware. Paradoxically, you may do this mostly by using software – circuit simulators, ECAD tools, Hardware Description Languages – but in some programmes you may even find yourself with a soldering iron in your hand.

Further reading

Clements, A. (2000) *The Principles of Computer Hardware*, Oxford: Oxford University Press.

Lunn, C. (1996) *The Essence of Analog Electronics*, London: Prentice Hall.

NOTES

1. John von Neumann was responsible for the design of the *Eniac* computer. For years it was believed that this was the first example of an electronic computer, but in fact it

had been preceded by *Colossus*, a British computer. The problem was that Colossus was developed in secrecy, during wartime, and its existence was an official secret for many years.

2. At least, at the time of writing, the currency is megahertz. If when you are reading this you would expect to read ratings in (perhaps) gigahertz, then you can be assured that **Moore's Law** has continued to hold true.

3. In fact, it is often more appropriate to measure performance empirically. For that purpose, certain standard programs, known as *benchmarks*, have been devised. These can be run on different systems and their completion times measured.

4. The low cost to the customer of buying a chip is only possible because, once the tools have been set up, they can be turned out in very large numbers.

4 COMMUNICATION AND INTERACTION

4.1 COMPUTER COMMUNICATIONS

The term *Information and Communication Technology* (**ICT**) was coined in recognition of the fact that what were once thought of as two separate forms of technology have largely merged. Successful communication depends on many components and factors. There must be some medium of communication between computers, and growing numbers of technologies are being developed. Physical forms of connection include wires and fibre optics, but increasingly *wireless* media are being used, including radio signals (such as Bluetooth) and infra-red light.

When a link has been established, then software has to manage the communication over it. This relies on the application of standards. In other words, two computers cannot share information and unless they both agree as to the format of that information and what the different components of it signify. For instance, a basic communication will generally include an address – its intended destination – and a message. The sending and receiving computers must agree which part of the communication contains which information and what to do with them.

No communication medium is perfect. Distortions and errors can be introduced into a message as it is transmitted. **Digital** communications are – or can be – relatively robust in respect of such errors. Extra information can be added to a message making it possible to detect the occurrence of an error. One method is for a computer sending a message to send also the number of characters in that message. If the receiver receives a different number of characters, then an error has occurred. By adding still more information with the message, it can be possible to pinpoint the error and even to correct it.

Much of the power of modern computer-based systems comes from the fact that computers are linked and can communicate. The next section describes how this happens, in practical terms, in computer networks.

Further reading

Ziemer, R. E. and Tranter, W. H. (1995) *Principles of Communications: Systems, Modulation and Noise*, Boston: Houghton Mifflin.

4.2   COMPUTER NETWORKS

One of the biggest changes in computing, in the past thirty years or so, is the Internet. The name 'Internet' refers to the fact that it is a set of networks which are all interconnected. Networks facilitate communication between people (email), but also have a number of other advantages: they make it possible to share resources. For instance, all the people in an office with their own PCs might share a single printer. Networks also increase reliability. If that same office has two printers and one breaks down, all the users will still have access to the second one (though the print queues on it may become longer).

Computers on a network will be connected in different ways. Different forms of connection run at different speeds and have different associated costs. The faster connections tend to be more expensive. The distances bridged also affect the choice of connection. The short distances of an office can be spanned quite quickly and cheaply with simple cables, while intercontinental connections might require a satellite link.

The shape or *topology* of a network is important. An obvious topology is to connect every computer directly to every other one. Such a network is said to be *fully connected*.

The advantage of a fully-connected network is that it will be highly reliable: if one computer breaks down, it does not affect communication between any others. However, fully-connected networks are generally impractical because of the cost of the connections, which grows exponentially. Most networks therefore have some level of partial connection. This means that it is generally possible to communicate from any computer to any other, but usually only through inter-mediate computers. In a partially-connected network, there will generally be more than one route that a message can take from one computer to another. Working out the route that messages should take is an important consideration. The objective is to use the network as efficiently as possible, to spread out the communication to minimise congestion in any part of it.

Sorting out the **hardware** to connect computers into a network is not sufficient; software has to be provided that will manage the communication. An important aspect of this is that the software has to know the format of the information that will be communicated. Thus there are rules (known as *protocols*) which are standardised across the network. These protocols can be quite complex and are often struc-tured in layers, reflecting the different levels of communica-tion that underlie them. Different protocols provide different features – different *qualities of service* – depending on the type of communication required. For instance, for some high-reliability purposes it can be important to ensure that a com-munication sent from one computer does arrive at its intended destination. When it does arrive, the recipient should send back an acknowledgement message. Of course, the picture is not as simple as implied. How long should the sending com-puter wait for an acknowledgement? If it has not received one, then it may be that the original communication has just taken a long time to traverse the network. Alternatively, perhaps the original communication did arrive at the destination and it is the acknowledgement that has got lost? Design of communi-cation protocols thus generally involves trade-offs between the speed and reliability required of the communication.

Communicating across a network, via intermediate computers, means that information can be intercepted. This raises questions relating to security. Encryption is a means of scrambling information in such a way that it can only be unscrambled, and therefore read, by the intended recipient. Another increasingly important security feature is the *firewall*. The name comes from the firewalls that are found in cars, forming a barrier between the engine and the passenger cabin. They keep bad things, such as heat and exhaust out of the car, but have holes that admit the necessary connections, such as the steering column and brakes. Computer firewalls have a similar role of keeping the bad stuff out while allowing the good things through. They may be implemented as hardware or software; a truly secure system will use both. The bad elements that they keep out include probes from other computer users who are looking to see whether you are running a system with security weaknesses that they can exploit. They also monitor suspicious activities on your computer and try to stop you inadvertently passing on any **viruses** with which your computer has been infected. At the same time, the firewall will let through the things you want: the emails and webpages that you want to see.

Networks have grown greatly in importance and continue to do so. Wireless connections (using radio, infra-red and similar communications technologies) increase the number and variety of devices connected. Increasingly, devices that would not normally be recognised as computers are becoming incorporated into networks: telephones, PDAs, clothes and so on. These developments will make networking technology even more important.

Further reading

Sloman, M. and Kramer, J. (1987) *Distributed Systems and Computer Networks*, New York: Prentice Hall.
Tanenbaum, A., Day, W. and Waller, S. (2002) *Computer Networks*, Englewood Cliffs, NJ: Prentice Hall.

4.3 CONCURRENCY AND PARALLELISM

It is often necessary or desirable to have more than one computer program (or, more strictly, *process*) running at the same time on some computer system (you can think of a process as being a program that runs on a *processor*; see Section 3.1: Computer Architecture). Such multiple processes running at the same time are said to be running *concurrently* or *in parallel*. Some systems have more than one processor, so can achieve parallelism by having different processes running on different processors, but even in a system with a single processor it is possible to have more than one process active at any time. Even though only one process can be physically running on the processor **hardware**, that processor can be shared between a number of processes. Each process will occupy the processor for a short time, with others waiting in the memory ready to have a turn on the processor. The time periods on the processor are so short and the swaps between processes so fast that to the human user it appears as if all of the processes are running at the same time.

There are a number of reasons why concurrency is a useful facility:

To fully utilise the processor
Processors run at speeds much faster then the input and output devices with which they interact. Remember that processors can execute millions of operations per second, but think how long it takes to print a single page on a printer. Imagine if the processor sat idle doing nothing until it was notified that the printer had finished. That would be a waste of billions of processor cycles.

To allow more than one processor to solve a problem
This is almost the obverse of the above case. You may have a difficult problem that requires lots of processing power to solve. If you can divide it up and solve little bits of it on different processors, then you may achieve a great acceleration in its solution. (This form of parallelism is also relevant to the topic of Distributed Computer Systems; see Section 4.5.)

To cope with parallelism in the real world

For instance, an airline may have one database containing details of its flights, but that database can be accessed by hundreds of travel agents and on-line web-based customers. They cannot each wait until they receive some kind of signal to tell them that the database is available for their exclusive access; they will be generating queries and updates in parallel at any time. Or, to take another example, an air traffic control system has to cope with the fact that there are lots of aircraft in the air all at the same time. Its essential job is to manage them all at the same time.

In cases such as these, the system must be able to accept inputs from different sources at the same time, and it will be designed to do this in such a way that the users are not generally aware of the other interactions.

To some extent parallel processes have to be independent of each other – otherwise they could not run at the same time. However, it cannot be the case that they are entirely independent and from time to time they will have to interact. This is what makes programming parallel processes quite difficult.

There are a number of particular problems that can occur in a concurrent system, including:

Deadlock

This is where one process (call it process P_1) cannot continue processing until another one (P_2) has finished what it is doing, but it turns out that P_2 cannot proceed until P_1 has got further. (Think of two men trying to get through a narrow door, and insisting 'After you, Clarence,' 'No, after you, Claude'.)

Interference

This is where two processes attempt to update the same resource at the same time. For instance, one travel agent books the last seat on the airline at precisely the same moment as a web user. Who gets the seat? Or does the aircraft end up over-booked with two passengers with the same seat allocation?

Concurrent systems and programming languages have been developed with facilities to assist in the avoidance of these problems. There are synchronisation primitives whereby one process may only be allowed to proceed when given 'permission' by another, and that permission will only be granted when it is safe to do so, when continuation will not lead to any of the problems outlined. These kinds of facilities may be built in to the operating system or they may be part of the programming language. In that case, it becomes the programmer's responsibility to write the code in such a way that errors do not occur. This can be difficult because interactions between processes and with the real world can be unpredictable in terms of the time that they take. This may mean that one set of processes runs correctly once, but the identical set generates an error (for example, a deadlock) when run again.

Concurrency and parallelism are examples of features that are frequently present on computer systems, but not apparent to most of the users. It is only thanks to the efforts of the system's designers that the system works in this way, securely and invisibly.

Further reading

Wellings, A. (2004) *Concurrent and Real-Time Programming in Java*, Hoboken, NJ: Wiley.

4.4 INFORMATION RETRIEVAL

In some contexts it is important to draw a distinction between *information* and *data*. The distinction is important in discussing information retrieval and databases. The things that are generally referred to as information are less specific, more fuzzy than data. Correspondingly, retrieving information is a less precise activity. This is not necessarily a bad thing. The user of an information system may not have an exact idea of what it is that they want to know anyway, so that a broad search may encompass whatever it is that they hope to find out.

The Web is the commonest (and largest) example of an information repository, and search engines are the most important information retrieval tool on the Web. Search engine queries are much less precise than database queries. Most queries to search engines consist of a set of words. The search engine will locate pages which contain all of those words. The common experience is that a large number of pages will be located in this way. It is possible to refine the query in a search engine to reduce the number of hits. For instance, the query

```
Computer AND Science OR Studies
```

would retrieve pages relating to Computer Science and Computer Studies. Notice that although this query may resemble an SQL query from Section 7.2: Databases, it is not going to be as precise in its retrieval; webpages lack the structure and logic of a database.

The less precise nature of information means that different techniques can be used in order to help users to locate the information they need. One example is the use of *case-based reasoning*. It is sometimes possible to abstract information so that it represents a characteristic *case*. Subsequently, someone else may need information that may be on a different topic, but that represents a similar case. *Case-based reasoning* is discussed below as a topic within Intelligent Information Systems Technologies, but might also be studied as part of information retrieval, given that cases are a particular form of information that one might wish to retrieve.

For the purposes of defining the topic areas for different taught modules, it is useful to be able to draw the kinds of distinctions made above between data, information and knowledge. In practice, though, the distinctions are not always adhered to, and there will be much overlap and cross-reference between the relevant topics as they might be taught in any degree programme.

Further reading

Baeza-Yates, R. and Ribeiro-Neto, B. (1999) *Modern Information Retrieval*, Reading, MA: Addison-Wesley.

4.5 DISTRIBUTED COMPUTER SYSTEMS

The conventional computer architecture in Figure 3.1 includes only one instance of each component, but there is no reason why a computer system should be confined to this form. Given the relative cheapness of components, and the availability of communications **hardware**, it is often more economical to have several processors and/or memory components connected together. This means that different parts of a computation can be carried out on different processors. In other words, the *Communications* component of Figure 3.1 can include communication with other computers in a network. When a lot of computers are connected together, they can share their resources (processors, memory and so on) and then various tasks can be completed with greater efficiency than they would be on a non-distributed system – a single computer.

As with many of the topics in this section, there are no agreed, hard-and-fast definitions as to what is a distributed computer system, and there are blurred boundaries between what would be taught in a module on that topic and what might be in other modules. As already implied, distributed computers are usually part of a network, so that topics such as inter-computer communications might be considered to be part of Distributed Computer Systems or part of Computer Networks.

The Internet represents one form of distributed system. It consists of millions of computers, so that each one has a processor and memory and is linked to others by the communications network. Nevertheless, the Internet can be used as a means of providing distributed processing. For instance, the Search for Extraterrestrial Intelligence (SETI, http://setiathome.ssl.berkeley.edu/) is running a massively distributed system across the Internet. When not in use, most computers sit idly doing nothing or displaying a screensaver. The SETI software uses the available processing power during these idle periods to do something useful, to search through vast amounts of data gathered from radio telescopes for patterns. Any such patterns that are found might indicate that someone or something out there is trying to contact us. It is possible to run this program in this distributed way because of the nature of the task. Each computer can be given data to analyse which is almost completely independent of the data being analysed on every other computer.

Having multiple processors available should mean that problems can be completed more quickly. You might assume that a system consisting of 1,000 processors could execute a program 1,000 times more quickly than a single processor. Unfortunately, that is not generally true. Unlike the SETI problem, most programs do not consist of completely independent components; there must be coordination and communication between the parts of the problem being tackled on each processor. This represents an *overhead*, an extra cost to the computation. At the same time, the 1,000 processors are only going to be useful if the particular problem can be broken down into 1,000 component sub-problems. A number of techniques and technologies have been developed to facilitate the sharing of (sub-)tasks among distributed computers, in such a way as to optimise performance and to achieve as much of the potential speed-up as possible.

There are a number of important topics within distributed computing, including:

Naming

Assigning names to objects may seem a trivial task, but in fact it is a fundamental operation in many contexts, not least in distributed systems. If a user needs to use a resource on a particular computer on the Internet, then they must have a way of specifying (naming) that specific computer – and no other one out of the millions that are on the net.

Sharing

As implied above, one of the positive features of a distributed system is the ability that it affords to share resources. This may be processing power, a printer, or a database, for instance. Managing such sharing can be quite complex. A printer can only print one item at a time, so if it is already printing one file, what happens if another user sends another file to be printed?

Availability and reliability

The components of a network are as prone as any other computer technology to failure. What effect does a failure have on the overall system? One of the advantages of distributed systems is that they are generally less vulnerable to component failures. This is because they usually include multiple (redundant) copies of components. For instance, if a printer is off-line, then a job that has been sent to it might be diverted to an alternative similar printer. There may be a degradation in performance (the second printer's print queue will get longer and each job take longer to complete), but the job will be completed.

On the other hand, some components of a distributed system may be critical and not duplicated. If such a component fails then the whole system may crash. It is up to the designer of the system to balance the need for reliability against cost.

Replication

In a distributed system there are often copies of information. A distributed database will not generally reside on a single computer, for example. The cost (in time and money) of accessing it from computers distributed around the world would be too great. Thus, parts of the database are likely to be copied on different computers. Naturally, this introduces complexity to the situation. It is necessary to keep all copies of information consistent. If two copies of the same information exist, which is right? For instance, what if one record on an airline database says that a particular seat on a flight is available, but another says that it has been booked?

Concurrency and synchronisation

If the subtasks of a computation are distributed across a system then there is a need to coordinate them. When the results of subtask S_1 become available, they will have to be combined with the results of subtask S_2, but if S_2 has not completed, then there will have to be a delay before the two are combined.

Time and coordination

Distributed systems inevitably involve time delays, not least because components are geographically dispersed and messages take time to pass between them. Nevertheless, it is often important that events occur in a particular order. In a bulletin board application, it would cause confusion if the response to a posting became available to some users before they had seen the original posting.

In other contexts, we often use time (that is, clocks) to facilitate coordination. It is not easy to use clocks in distributed systems. Whereas a difference of five minutes in the time shown by my watch and yours might cause some frustration if we arrange a meeting, such differences (and much smaller ones) might cause chaos in a distributed system. You and I might try to avoid the problem by synchronising to an agreed standard time (for example, the Greenwich Time Signal), but it is not so easy for computers to do that to the accuracy required. By the time any time signal has reached your computer, it is already out of date, and if the delay in reaching my computer was greater, then we again have a problem. Fortunately, a number of techniques have been developed to deal with these problems in distributed systems.

The importance of **Moore's Law** is mentioned often in this book. Moore formulated it in terms of hardware, the number of transistors on a chip, but as well as making processors more powerful, advancements have been made in making better use of those processors. Distributed computing is a way of doing this, and it will become increasingly important in the development of Computer Science.

Further reading

Coulouris, G., Dollimore, J. and Kindberg, T. (2005) *Distributed Systems: Concepts and Design*, Reading, MA: Addison-Wesley.

Crichlow, J. M. (2000) *The Essence of Distributed Systems*, Harlow: Pearson Education.

Sloman, M. and Kramer, J. (1987) *Distributed Systems and Computer Networks*, New York: Prentice Hall.

4.6 DOCUMENT PROCESSING

There was a time when technologists looked forward to the *paperless office*, when the use of computer technologies would make the use of paper obsolete. As anyone who has been in a hi-tech environment knows, if anything the obverse is true: computers seem to generate *more* paper in the form of various print-outs. Looking at this in a positive light, the technology has enabled us to produce not just more paper but also better quality, attractive publications.

There are two broad approaches to document production: *word processing* and *text processing*. Word processing is based on the interactive generation of a document which looks on screen the same as it will look when printed on paper. This is known as *what you see is what you get* or **wysiwyg** (pronounced 'wizzywig'). This is the approach that any user of Microsoft Word or any similar word processor will be familiar with. Such tools make it easy to produce documents that make full use of the flexibility of modern printing. Or, to put it another way, they make it easy to produce ugly documents which give the reader the impression that the producer of the document was trying to show off. Any document with more than three different fonts on one page should be treated with suspicion. Modern information technology has replaced the traditional printing technologies, but the established arts of design and typesetting are as important as ever.

Modern word processors do much more than the typewriters that they replaced. They will check your spelling and give you hints on grammar, for instance. The technology behind such tools is very interesting in its own right. A spelling checker does not simply look up every word in a massive dictionary, for instance. A dictionary that was big enough to contain all forms of all words (for example, all the possible tenses of each verb) would be too big and too slow to search. A dictionary *is* used, but if a word occurs that is not in the dictionary, but is similar to one that is, then a set of rules is applied to see whether the dictionary word can be transformed to the one in the text. The simplest example is the plural, so that if, say, the word 'elephants' is in the text but is not found in the dictionary, it is matched with the entry 'elephant' in the dictionary. Then the spelling checker will apply one of its rules for making plurals and hence accept 'elephants' as a correct spelling. Documents are not purely text either. They usually contain graphics and other forms such as tables. Again, the tools and interfaces for designing and generating these can be interesting in their own right. Other traditional phases of document production that have been facilitated by technology include the generation of tables of contents and indexes. Creating indexes, in particular, used to be a tedious job for a human worker.

The other approach to document generation, text processing, works in a rather different way. This is a two- (or more) stage process. First the author produces a text document that is *marked up* with formatting instructions, then this document is passed through software which converts it into the printable form, with the formatting instructions put into effect. There are a number of different text processing tools. The most used is Latex (pronounced 'Lay-tek'), which is a development of Tex ('Tek'). An example of Latex is that the following text is marked up with the commands or *tags* \textnormal and \emph ('emphasis'). Thus,

\textnormal{Sometimes \emph{we \emph{discover} unpleasant} truths.}

would appear in the final, formatted document as:

Sometimes *we* discover *unpleasant* truths.

Text processing is most useful for scientific writers who find that word processors cannot produce the kind of output they need. In particular, Latex has extensive facilities to typeset very complex mathematical material. There are also facilities available to produce other specialist output, such as chemistry and music. Another advantage of the text processing approach is that it allows a degree of programming-like facilities. For instance, subsection headings may be automatically numbered. The software will automatically keep track of the numbering and assign the appropriate number to each heading. If the writer decides to insert a new subsection in the middle of the existing ones, he or she does not have to worry about disturbing the sequence of numbers; the software will do it for them.

There are, however, a number of problems in using this kind of software. An obvious one is that the writer has to get the mark-up right. In other words, if the writer were to misspell a tag (for example, \emhp for \emph), then the Latex processor will not recognise the tag and it will be unable to process it. At best, Latex will generate an error message and the writer will have to go back to the marked-up document and fix the error. At worst, the document will be printed but will come out not looking as the writer intended. Another problem is that Latex is not well standardised. Particularly when using specialised notations, a writer may have used a particular set of Latex definitions, or *libraries*. If a colleague receives a copy of the marked-up text, but has a different set of libraries, then he or she may not be able to print the same document.[1]

The advent of new technologies has also had wider-ranging implications. In particular, the legal framework surrounding the copyright of documents has to cope with new challenges. Once a document is available in an electronic form, it can be reproduced in large numbers for minimal cost. For this reason, publishers are quite wary of using electronic documents. Copyright law and its application in an electronic age is a very important topic.

Further reading

Ransom, R. (1987) *Text and Document Processing in Science and Technology*, Wilmslow: Sigma Press.

4.7 GRAPHICS AND SOUND

Computer graphics literally represent one of the most visible developments in Computer Science, yet at the same time their improvement might be said to make them *less* apparent. If you go to the cinema and sit right through to the end of the final credits, then in almost any current film there will be a credit for the graphics programmers. This applies not just to films such as the *Toy Story* and *Shrek* series, which consist entirely of computer-generated animation, but to many other mainstream films and action dramas. Should this come as a surprise to you at the end of any film, then the programmers have achieved their objective: they have synthesised pictures that are indistinguishable from cinematic photography.

Films are perhaps the most glamorous form of computer graphics, but nearly all modern computer systems have interfaces such as Microsoft's Windows, which are known as graphical user interfaces (GUIs). As implied by the name, these rely on computer graphics – to generate the icons, buttons, windows and all the visual elements. In between these extremes are a large number of applications for graphics, including diagrams and business graphics, cartography, medical imaging and computer-aided design (CAD).

In computer graphics, the screen display is treated rather like a sheet of graph paper. Just as you can number the squares on a graph, any point on the screen can be represented as a pair of numbers, x and y. The squares of the graph paper correspond to *picture elements* or **pixels** on the screen. There is a number associated with each pixel which specifies its status. On a simple black-and-white display, that number can be a simple binary digit (*bit*) where a 1 means that the pixel is black and a 0 means it is white. A colour display needs a longer number to represent each pixel – representing the relative amounts of red, green and blue that make up the colour of the pixel.

There are two levels at which you can tackle the generation of computer graphics. One is to use a graphics package. This is essentially a library of program procedures that will draw and manipulate graphical objects. For instance, there will be procedures that will draw lines, rectangles and circles of specified sizes, positions and colours. For example:

```
draw_line(x1, y1, x2, y2)
```

might cause a line to be drawn between the points $(x1, y1)$ and $(x2, y2)$ and

```
draw_rect(x1, y1, x2, y2, colour)
```

could draw a rectangle in colour *colour* with its bottom-left corner at $(x1, y1)$ and its top-right corner at $(x2, y2)$.

A program consists essentially of a set of calls to these procedures that will draw the specified objects to make up whatever image the programmer requires. Pictures consist of a set of shapes or *polygons* composed together. For example, the picture of the house in Figure 4.1a is composed of the four rectangle and one triangle polygons, as in Figure 4.1b.

Use of such graphics packages has a number of advantages. For a start, it means that the programmer does not have to be concerned with many of the low-level operations involved. At the same time these packages are implemented on different systems. This means that the programmer does not have to be concerned about differences such as the type of display to be used. The same program should produce the same graphical output on, say, a Macintosh laptop's LCD screen as on a

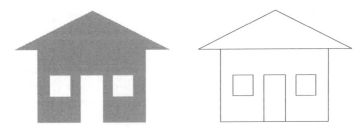

Figure 4.1 A graphic, composed of polygons. (a) shows the picture as it would appear to the viewer, a picture of a house, but (b) shows that underlying it is a simple set of coloured polygons, in this case four rectangles and one triangle.

large-scale data projector driven by a Windows PC. Many standardised graphics packages are available, as certified by the American National Standards Institute (ANSI) or the International Organization for Standardization (ISO).

The other level of graphics programming is precisely where the programmer *is* concerned with the low-level details. An obvious example would be the programmer who implements any one of the graphics packages. This level of programming relies on a lot of mathematics. Geometry and trigonometry are important. For instance, a package might include a procedure

```
draw_circle(centre, radius, colour)
```

which will draw a circle with the given radius, centred on the specified point. The implementation of the draw_circle procedure will have to calculate which pixels need to be changed to *colour* to represent the circle. Vectors and matrices are a convenient way of representing graphical objects and their transformation.

Graphics displays (screens, paper and so on) are two dimensional, but the pictures we want to display on them are often of three-dimensional scenes, which further complicates the manipulations. The program can calculate where the objects are in a three-dimensional space and then project them onto a two-dimensional plane. This can be quite complex. For instance, parts of a solid shape can hide other elements which are behind them. The graphics software therefore has to calculate which lines and surfaces will be visible to the viewer and draw only those.

Graphics are often combined with sounds under the heading of *Multimedia*. The sound generation capabilities of PCs and other computers have improved with the provision of sound cards as standard. However, it has to be said that the use of sounds has not developed to the same extent as visual representations. Apart from MP3 music files, the most complex sound played by most computers is a set of simple beeps when an error occurs. Nevertheless, a great amount of effort has been expended on finding better and more adventurous ways of using sounds in interactions and these are summarised in Figure 4.2. There is little doubt that the use of sounds in computer interfaces will continue to grow in importance in time.

Speech
       Recorded (digitised)
       Synthesized

Non-speech
       Music
       Symbolic
              Earcons
       Figurative
              Auditory icons
       Analogous
              Sonification

Figure 4.2 Characterisation of the sounds that can be used in human-computer interactions.

Speech has been recorded in an electronic form on tape recorders, but it can now also be recorded digitally, so that it can be manipulated using computer technology. It has the limitation that only those words or phrases that have been recorded can be replayed. In contrast, arbitrary utterances can be generated by using synthetic speech. This will take a string of text and then generate the corresponding sounds of speech. The test of quality is how similar the speech is to that of a person. The best known example is probably the voice of Stephen Hawking's communication device, but it does use quite old technology and much better synthesisers are available. The advantage of a speech synthesiser is that it will speak *any* text presented to it. (Clearly, it is vital that the communication device used by a theoretical physicist should not have a limited vocabulary!)

Non-speech sounds can also be used to convey large amounts of information. Music is clearly one form of non-speech sound, and much computer technology has been devoted to the creation, storage and playback of music (think of CDs, MP3 players and so on).

Music-like sounds can also be used to convey information. Many computer programs use simple sounds, such as 'beeps' to signal errors, but they can also use more complex sounds to convey more information. Arbitrary sounds can be combined together in a structured way to create so-called *earcons* (a pun on the visual 'icon'). Although these sounds are artificial, people can learn what their meanings are. Another way of using non-speech sounds is to use recordings of sounds which resemble the object that they represent. These are known as *auditory icons*.

Although the technology for playing and manipulating sounds has existed for some time as a standard part of most PCs, the use of sounds has been somewhat limited. This is clearly an area in which there is scope for much further development.

Further reading

Cooley, P. (2000) *The Essence of Computer Graphics*, London: Prentice Hall.

Edwards, A. D. N. (1991) *Speech Synthesis: Technology for Disabled People*, London: Paul Chapman.

Foley, J. D., van Dam, A., Feiner, S. K., Hughes, J. F. and
   Phillips, R. L. (1994) *Introduction to Computer Graphics*,
   Reading, MA: Addison-Wesley.
Gaver, W. W. (1997) 'Auditory interfaces', in M. G. Helander,
   T. K. Landauer and P. Prabhu (eds), *Handbook of Human-
   Computer Interaction*, Amsterdam: Elsevier Science.
Pitt, I. and Edwards, A. (2002) *Design of Speech-based
   Devices: A Practical Guide*, London: Springer.

4.8   OPERATING SYSTEMS

The traditional configuration of computer **software** and **hardware** is as shown in Figure 4.3. Underlying the whole system
is the hardware, the processors, memory, discs and so on. At
the top level are the applications programs: the word processors, spreadsheets and the like. It would be difficult for every
application program to be written to interact directly with the
hardware: there are differences in the way different hardware

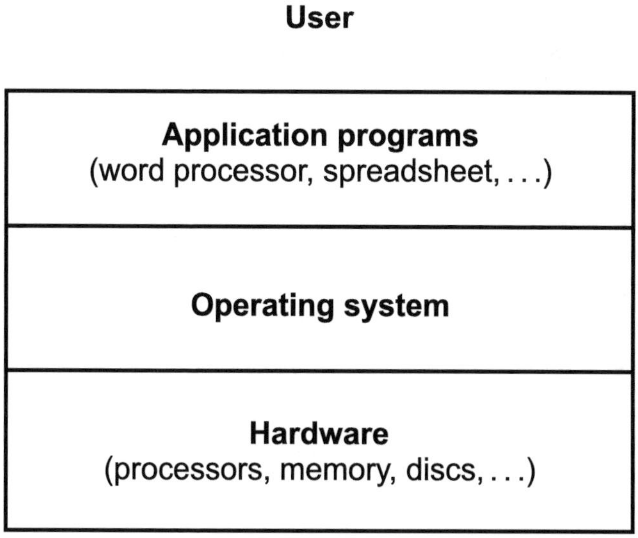

User

Application programs (word processor, spreadsheet, . . .)
Operating system
Hardware (processors, memory, discs, . . .)

Figure 4.3 The conventional relationship between hardware and
software. At the bottom is the hardware: processors, discs, and so on.

operates. For instance, two discs may have the same capacity, but require completely different commands to access them. Therefore, there is always an additional piece of software between the applications and the hardware. This is the *operating system*.

The operating system provides a uniform interface to the applications software. For instance, if a program is to write some data to a disc file, it will issue *system calls*, and *drivers* in the operating system will translate these into the appropriate low-level commands to the hardware discs. The system calls will be identical, whereas different drivers will be provided for different makes and models of disc.

The most apparent part of the operating system to the user is the user interface. For instance, the interface provided by Microsoft Windows is different to that of the Macintosh OS X, but they both provide the same basic facilities, such as folders, file manipulating, printing and so on.

The operating system makes it possible to run several programs at the same time. This is necessary when more than one person is using the same computer at a time (as in the case of mainframe computers) but is also a necessary part of using a personal computer. Indeed, the operating system itself is a program that has to run in parallel with any applications. Central to this sharing of the computer between programs is the concept of the *process*. Generally, each program appears to the operating system as a process and it manages the sharing of the hardware resources (processors, memory, communications and so on) between the processes. Thus, the operating system might allow one process to execute on a processor for a short time but then stop it to allow another process some processor time. Processors and memory operate so quickly (in human terms) that such swapping occurs too quickly to be observed by the computer user, and it appears to them that all the processes are running at the same time – even though there may in truth be just one processor and many processes.

The operating system ensures fairness of sharing of resources between processes, with the objective of as many processes completing their function in the minimum time. An additional obligation on the operating system is to enforce security restrictions. That is to say that where different processes owned by different users are sharing hardware, there is the possibility that they will interfere with each other, either deliberatively or inadvertently. For instance, one process which is executing a program that has a bug in it might attempt to change the contents of an area of memory that has been allocated to another process. The operating system will block that access, probably causing the offending process to fail (that is, crash). If a system is not completely proof against such accidental interactions, it will also be vulnerable to malicious attacks.

Most computers are connected in networks, and ultimately connected to most other computers via the Internet. Thus, operating systems have also to handle communications as part of their role. The interconnection of computers makes them more vulnerable to security breaches because people can access computers to which they have no right of access. Such attacks are generally gathered under the collective description of **viruses**, though strictly speaking a virus is just one means of attack. Modern operating systems thus have to be written so that they will manage all of the resources on the computer, and, at the same time, be proof against malicious attempts to disrupt those resources.

Further reading

Tanenbaum, A. S. (2001) *Modern Operating Systems*, Englewood Cliffs: Prentice Hall.

4.9 MIDDLEWARE

Figure 4.3 depicts the traditional configuration of application, operating system and hardware. The operating system serves as an interface between application programs and the computer hardware. For instance, to you as a user, any Windows PC looks essentially the same as any other Windows PC. In fact, the underlying hardware may be different – components from different manufactures are used in different configurations – although they are all quite similar. Larger computer systems – and particularly networks of computers – will vary to a much greater extent. Different computers on the same network will have different architectures, have different facilities (for example, printers) and even be running different operating systems. It is still a good idea, though, if applications can run anywhere on the network. This is the purpose of *middleware*, which presents a uniform interface to applications across networks, as illustrated in Figure 4.4.

User

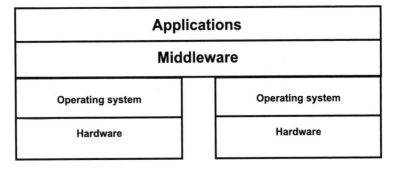

Figure 4.4 Middleware forms an interface between different computers on a network. The figure shows just two computers, but in practice there are likely to be many more, different ones.

That any company will run a variety of computers in different roles, in different departments, is almost inevitable. This is often historical. For example, a lot of companies have grown through acquisition, so that different parts of the one company used to be different companies. Each of those component companies had gone through their own IT evolution, and are probably geographically dispersed. For the new, amalgamated company to work efficiently, it is vital that the different systems be integrated, and that is the role of middleware.

One style of programming is *object oriented* (see Section 2.2: Comparative Programming Languages). Such programs are composed of *objects*. Programmers will usually write some of the objects themselves, but they will also use some that have been programmed by other people and collected into *libraries*. One of the strengths of object oriented programming over other styles is that libraries of objects are particularly easy and useful to share, and these present a mechanism for providing the functions of middleware. There are two reasons for using object oriented middleware. Firstly, a lot of modern software is written in that style, so that the systems with which the middleware is interacting may well be object oriented. Secondly, object middleware is more flexible.

The degree of heterogeneity of computer systems is probably increasing as a result of technical and commercial factors. Hence, the importance of middleware is growing correspondingly and the techniques used are likely to develop and improve.

Further reading

This is a relatively new topic. One good introductory text is:
Britton, C. and Bye, P. (2004) *IT Architectures and Middleware*, Boston: Addison-Wesley.
A useful, brief introduction is provided in:
Bernstein, P. A. (1996) 'Middleware: a model for distributed system services'. *Communications of the ACM* 39 (2): 86–98.

4.10   NATURAL LANGUAGE COMPUTING

As discussed elsewhere in this book, computer programming must be carried out in special, artificial languages, which have been specifically designed to avoid the ambiguity that is inherent in *natural* languages (such as English or French). Nevertheless, it is often useful to have computer software that can handle some aspects of natural languages. There are a number of applications in which it is desirable for a computer to have some level of 'understanding' of natural language. That is to say that the user can generate input in a language such as English and the program will be able to extract some form of meaning out of it. Ideally that meaning should be the same as that which a human listener would have extracted.

The input might be typed on a keyboard, but even more natural is if the user can speak to the computer. *Speech recognition* consists of translating the sounds of speech into their textual representation (as was once undertaken by secretaries taking dictation). A textual input – from a keyboard or a speech recogniser – can then be processed in an attempt to extract its meaning. Speech recognition software has become increasingly reliable in generating the actual words spoken, but it is still less than 100 per cent accurate. Still, this is a relatively easy problem compared to natural language understanding.

A certain amount can be achieved by simple *keyword spotting*. Particular words in the system's dictionary are searched for in the input and matched to appropriate responses. Words not in the dictionary are simply ignored. Of course, this can lead to misunderstanding because information has been lost: the true context of the recognised words. Ideally, that context should be structured in terms of the *grammar* of the language. A grammar can capture some of the structure of a language. For instance, a simple rule of English is that a sentence must contain a finite verb. Furthermore, a common structure for a sentence is

```
sentence = subject verb object
```

Other rules might be

```
subject = 'a' OR 'the' noun
object = 'a' OR 'the' noun
```

A program might examine an input sentence, matching words in its dictionary against this rule and hence extract the meaning. For instance, if *boy* and *ball* are recorded in the dictionary as *nouns* and *hit* is a verb, then the system would be able to recognise the sentences

> The boy hit the ball.
> The ball hit the boy.

Furthermore, the program would recognise that there is a difference in the meaning between these sentences because of the different roles of the subject and the object in a sentence. This is the advantage of a grammar-based approach; a simple keyword-spotter that merely picked out the words *boy, ball* and *hit* would not be able to detect the difference between the two sentences.

While using this grammar helps, it is limited because the sentences of English do not fall into this structure. In fact, there is whole range of different sentence structures. It is not possible to capture them all in simple rules like those above, and yet the different structures convey different meanings. Different techniques and formalisms are being developed and they are becoming better at capturing ever-larger subsets of natural language utterances.

Inputting natural language to computers is not the only aspect of natural language processing – it is often useful to generate natural output. The simplest way is to merely reproduce pre-stored utterances. Rather more challenging is to take some abstract representation of meaning and (using some form of grammar) generate words that embody that meaning.

However generated, the words of an utterance might be displayed as text on a screen, but they can also be rendered in audible speech, using a *speech synthesiser*. Speech synthesisers have been developed which can be fed printed text and will output the words as clearly understandable spoken words. Most synthetic speech is clearly recognisable as being mechanically generated, but it is constantly improving, to the point that some synthetic speech is not easy to distinguish from human speech when heard over a telephone line.

The mismatch of people and machines is never more apparent than in questions of the languages that we use to communicate with them. Programming languages must be precise and unambiguous; natural languages are full of ambiguity. Nevertheless, real progress is being made at adapting computers to make them capable of using the kind of language that we find more comfortable.

Further reading

Edwards, A. D. N. (1991) *Speech Synthesis: Technology for Disabled People*, London: Paul Chapman.
Pitt, I. and Edwards, A. (2002) *Design of Speech-based Devices: A Practical Guide*, London: Springer.

4.11   MULTIMEDIA

Almost any kind of information can be represented in a **digital** form. In turn, that means that it can be processed, communicated and presented on computer equipment. These different forms of information are generally referred to as *media* and because there are a number of them, they are multimedia. Examples of the different media are text, graphics, photographs, video images and sound.

These media are typically very rich in information. Take video, for instance. Companies pay thousands of pounds for television adverts that last only a few seconds because they can convey a large amount of (compelling) information in the moving pictures and sounds of the advert. The richness of a video representation is reflected in the digital representation. That is to say, the amount of memory required to store the digital representation of, say, a few seconds of video is very large. While increased processor power and reduced cost of memory makes it more feasible to handle multimedia on computers, there is still a need to reduce the amount of memory required. Thus, a lot of work goes into compression techniques. An ideal compression method will reduce the amount of memory required to store multimedia data without degrading its quality.

Probably the best known technique currently is MP3, as used by music storage and player systems. This method makes use of the fact that there is information in a sound signal which is *redundant* – that is to say, some components of the sound cannot be heard by the human ear. They can thus be filtered out of the digital representation, reducing its size but maintaining its quality.

Compression is just one of the manipulations that can be carried out on multimedia data. While one objective of this kind of manipulation is to not distort the original signal, others will deliberately change it. Special effects may be added to a sound or video file, as demonstrated in almost every pop video.

Multimedia are increasingly becoming part of everyday computer interaction. Tools are available to assist in the creation and manipulation of multimedia materials. Yet there is still scope for finding better ways to interact with multimedia materials, and this is a continuing topic of research.

Further reading

Chapman, N. and Chapman, J. (2004) *Digital Multimedia*, Reading: John Wiley.

4.12   HUMAN-COMPUTER INTERACTION (HCI)

It has to be said that not every encounter between people and computers is positive. We can all list the frustrations we have experienced when the software has not done what we expected or when we cannot find the way to make it do what we want. These frustrations are often caused by the design of the interface between the user and the software: the *human-computer interface*. The problems are often due to the fact that the people who design the interface – the programmers – are not familiar with tools, techniques and principles that have been developed to improve interface design.

There are two sides to the interaction. On one side is the computer, which Computer Scientists understand well and is (relatively) predictable and *deterministic*. On the other side is the person, who is rather more complex, variable and unpredictable. In other words, the human side of the interface may belong more in the realms of psychology, and, indeed, HCI is often seen as a multi-disciplinary topic, involving at least the two fields of Computer Science and Psychology.

Sometimes the problems generated by human-computer interfaces are due to the fact that the designers have been more concerned with what is easy to implement on the computer than what users will find convenient. To counteract that, the concept of *user-centred design* has been developed and advocated, though there are different ideas as to how such orientation to the user can be achieved, and what *user-centred* means.

If the theoretical foundations of Computer Science are somewhat thin, there are even fewer laws that can be applied to the human side of the interface. What there are instead are *guidelines* – derived over years of study and experience. The most comprehensive set is probably that derived by Smith and Mosier (1984), but shorter summaries such as Shneiderman's Eight Golden Rules (Shneiderman 1998) are rather easier to handle:

1. Strive for consistency
2. Enable frequent user shortcuts
3. Offer informative feedback
4. Design dialogues to yield closure
5. Offer error prevention
6. Permit easy reversal of actions
7. Support internal locus of control
8. Reduce short-term memory load.

What these rules do is largely to distil knowledge of people's strengths and weaknesses in interaction. Designing an interface with such rules in mind is a good start towards creating a usable product. However, the principle of user-centredness implies that simply following rules is not enough, that people must be involved in the development and evaluation of interfaces. Thus, in the early stages of a design process, the needs of the eventual users must be elicited and taken into account. Techniques have been developed that can be applied to eliciting the required knowledge. These include focus groups, interviews and questionnaires. Prototyping can also be an important tool, whereby users get to try out something which looks like the proposed design, but is less than a complete implementation.

In a truly user-centred project, testing will be involved at all stages. For instance, if the classic *waterfall* engineering approach is followed, as illustrated in Figure 2.8, then users would be involved at each of the phases. However, in most cases, the most important stage to involve users is at the testing phase. That is to say, an (almost) final version of the system must be tested by users. A number of techniques have been developed which can be applied to evaluation.

For many years the human-computer interface has been embodied by the keyboard, screen and mouse, but research is constantly under way into the use of new forms of communication. For instance, there is no room on a PDA (personal data assistant) for such devices, so new forms of interaction must be developed, such as speech, non-speech sounds and handwriting recognition. Many other topics that might be studied under some of the other headings in this part, including Multimedia and Graphics and Sound, will also be relevant to the study of human-computer interaction.

Further reading

Dix, A., Finlay, J., Abowd, G. and Beale, R. (2003) *Human-Computer Interaction*, London: Prentice Hall.

Faulkner, X. (1998) *The Essence of Human-Computer Interaction*, London: Prentice Hall.

NOTE

1. Such incompatibilities are not completely unknown to the word processor user too. If the owner of the latest version of a word processing package uses its new features, then a colleague with an older version will not be able to see the use of those features.

5 PRACTICE

5.1 E-COMMERCE

As is well-known, the Web was originally envisioned by its inventor, Tim Berners-Lee, to facilitate communication between scientists. In its early days its use reflected these non-commercial, collaborative origins. People shared information and ideas freely and any apparent attempt to make money was often frowned upon. This situation did not last, as people came to realise there was money to be made, and the phenomenon of *E-commerce*[1] was created.

E-commerce is generally classified as *business to customer* (B2C), *business to business* (B2B), or *customer to customer* (C2C). B2C corresponds to retail selling, that is to say the kind of transaction that goes on between Amazon and its customers. B2B corresponds to wholesale trading, whereby businesses buy and sell their raw materials and products electronically. In conventional commerce, C2C might be represented by using classified ads. E-bay is the most popular example of C2C e-commerce.

E-commerce can further be classified as *hard* or *soft*. Soft e-commerce essentially implies that information only is exchanged. Thus, websites that offer information regarding companies and products are examples of soft e-commerce. Search engines, such as Google, represent an interesting example of soft (B2C) e-commerce. It costs nothing to perform a search and yet Google is undoubtedly a commercial operation (just look at how much its founders have earned!). Google makes money by displaying *sponsored links* – they are effectively adverts, which are paid for.

Hard e-commerce involves the exchange of money for goods or services. Amazon would be an obvious example of hard B2C e-commerce. With credit cards and other similar mechanisms, the payment can be made directly on-line. Software can be downloaded directly so that the whole transaction happens electronically, but other, 'hard' products inevitably require other agents to physically deliver the goods. There are any number of B2B hard e-commerce websites. The essential difference is that they provide the kinds of goods that businesses need.

All networks have security requirements and this is clearly important in (hard) e-commerce in which – in effect – money is being transmitted across the network. Entering one's credit card details on a web form is an interesting case study which has implications for many aspects of e-commerce.

In order to buy goods with your credit card, you must provide certain information, including the credit card number and its expiry date. *Anyone* with all that information can obtain money from your credit card company – not just the owner of the webpage displaying the particular item you want to buy. Therefore, while that information is in transit across the network, it is not sent as plain, readable text: instead, it should be scrambled or *encrypted*. Only the intended recipient will be able to unscramble it; anyone else intercepting the message will see only an apparently-random piece of text.

Of course, the intended recipient must use that information only for the purposes for which you have provided it. That is to say, they will charge you the cost of the goods you ordered, and no more. It is illegal for them to pass on any of the information you have provided – not just your credit card details but also your name and address and other details, unless you have explicitly given them permission to do so. At least that is illegal in the UK. One of the problems of the Internet is that it spans national boundaries. Even a company with a *.uk* address might not be registered in the UK, so is it bound by the same laws? These kinds of questions are arising with increased regularity and are likely to become increasingly common as e-commerce continues to expand.

One thing that an unscrupulous company might do with some of your data is to sell your email address to someone who collects addresses, so perpetuating the most annoying form of e-commerce: Spam.[2] This is another interesting example of the consequences of e-commerce. Legislation exists that theoretically makes spamming illegal, but it does not seem to have had any effect.

It is a simple rule of existence that if there is a way to make money out of something (honestly or otherwise) people will do so. That is how it has been with the Internet. Whatever the original motivation behind the invention of the Web, it is now the centre of many large businesses. This is likely to hold true for the foreseeable future, and new businesses, techniques and technologies will be developed to make the most of the opportunities.

Further reading

Deitel, H. M. and Deitel, P. (2000) *E-Business and E-Commerce: How to Program*, London: Prentice Hall.

5.2 PROFESSIONALISM

Any society is based on a system of ethics – rules by which the members of that society are expected to abide for the mutual benefit of the society's members. In other words, *ethics* define what is considered to be right and wrong. Any technology must be used within the accepted ethical codes and the more powerful the technology the more important that it is used in an ethical manner. However, information technology can raise particular ethical dilemmas.

If ethics define what is right and wrong, there are still questions of degree. If something is considered to be very wrong, then the society is likely to adopt a law against it. In other words, violation of that particular rule is discouraged by the threat of punishment. To take a simple example, theft is considered wrong in most – if not all – ethical codes. This is a rule embodied in law. A problem is, though, that the traditional concept of theft is where one person is deprived of the owner-ship and use of an object by another. What if the thief can take something without depriving the original owner of its use? This is precisely the situation with electronic software and data. Limitless numbers of copies of a music recording can be made without depriving anyone of the ability to listen to that tune.

Existing ethical codes – and corresponding laws – do not fit the new situation generated by the advent of information technology. Take another example of a computer program. If someone invests time and money in creating a piece of soft-ware, the prevailing ethical code suggests that they deserve reward for it, that their investment should be protected. But what is to be protected? Is it the idea behind the programme, the algorithm? If so, that is intellectual property which would normally be protected by way of a patent. But patents are notoriously difficult (and expensive) to obtain. It has to be proved that the new idea is sufficiently different from any prior inventions, that no one else had the idea first and so on. So, one might try to protect the software itself. In contrast to a patent, the copyright of a new artefact is freely and auto-matically granted to its creator. However, electronic copyright is hard to enforce. Who is going to know that the person working quietly in their bedroom is using a copy of Windows that they did not pay for?

So, it is hard for existing laws to cope with the new features of electronic technology. There have been some almost bizarre attempts to do this. For instance, in the 1990s the Apple Computer Company claimed that Microsoft's Windows oper-ating system was too similar to that on the Apple Macintosh. Apple could not claim that any program code from their system had been put into Windows – which would have been a violation of copyright – and so they sued Microsoft on the basis that they had stolen the 'look and feel' of their software. Given the vagueness of this term, it is probably not surprising that Apple lost the case.

These are just a few examples of the many new ethical uncertainties that are generated by the technology. The Internet raises many more. Ethics and laws vary from country to country. Given the distributed, global nature of the Internet, the laws of which country should be applied? If a user in country X views a webpage containing material that is illegal in that country, can any action be taken if the owner of the webpage is in country Y, where it is perfectly legal? Would it make any difference, if the webpage server is in country Z, where another law applies?

Data protection is another important area. If a company collects personal data for one purpose (for example, to deliver goods that a customer has ordered), they should not use it for another purpose (for example, marketing other goods to that person). Furthermore, there are dangers that they might combine the information that they have with that from other sources and be able to make inferences about that person.

It is in the world of such uncertainties that the computer professional has to operate. Professional associations have an important role to play in that they define codes of ethics by which their members must operate. The foremost such societies for UK Computer Scientists are the British Computer Society (BCS) and the Association for Computing Machinery (ACM).

While it is important preparation for a career in computing to study these topics at a 'theoretical' level, it is almost certain that any Computer Science student will be confronted with some of these ethical dilemmas during their studies. The student should be prepared to understand the implications of decisions they make. Plagiarism in various forms is a major concern for all university authorities. Again, there are particular dangers within Computer Science, where questions about the ownership of the intellectual rights to a piece of software are just as relevant to an assessed programming assignment as to a piece of commercial software, as discussed above. (See also Section 12.4: Coursework, in Part III.)

Further reading

Ayres, R. (1999) *The Essence of Professional Issues in Computing*, London: Prentice Hall.

Johnson, D. G. and Nissenbaum, H. (2004) *Computers, Ethics, and Social Values*, Englewood Cliffs, NJ: Prentice Hall.

5.3   SIMULATION AND MODELLING

Traditional science is based on carrying out experiments (with chemicals, mechanical objects, living plants or whatever). In some cases, however, it is desirable to carry out experiments but for one reason or another it is not possible to experiment with real, physical materials: it might be too dangerous (for example, experimenting with nuclear materials); the system to be tested may be unique, so that there is no 'spare copy' to manipulate (for example, the national economy) or the experiment may imply looking into the future and time-travel is not an option (for example, weather forecasting). In such cases it is possible to build a computer *model* of the system of interest, and then to conduct experiments on that model. In other words, one *simulates* the system.

A well-known economic model is illustrated in Figure 5.1. This is a model of *supply and demand*. Demand is represented by the dashed line going from top left to bottom right while supply is the solid line, rising to the right. The slope of the demand line indicates that a greater quantity will be demanded when the price is lower. On the other hand, the slope of the supply line tells us that as the price goes up, producers are willing to produce more goods. The point where these curves intersect (P, Q) is the *equilibrium point*. At a price of P, producers will be willing to supply Q units per period of time, and buyers will demand the same quantity.

The ideal is to be able to manufacture items at the rate Q and sell them for price P. The diagram also illustrates what would happen if the price was set too high, at P_1. Some customers would perceive that price as being too high and would not purchase the item, hence the quantity sold would be less (Q_d) but the amount manufactured would be greater (Q_s). In other words, there would be an over-supply ($Q_s - Q_d$). Clearly this is undesirable because it implies wasted goods. Hence,

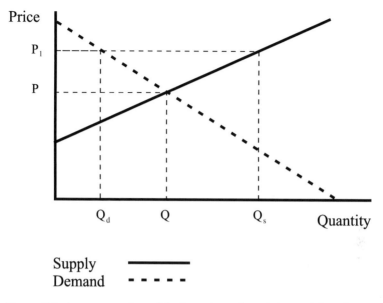

Figure 5.1 A representation of the law of supply and demand (based on http://www.answers.com/topic/supply-and-demand).

it would be good to be able to model the system and thereby calculate the appropriate values for P and Q.

Figure 5.1 represents such a model, and it would be relatively easy to implement this model on a computer. Unfortunately, though, there are questions open about the model's exact configuration. What are the slopes and intercepts of the supply and demand lines? Indeed, should they be straight lines at all? Might they not be curved? Also, the model may be an over-simplification. For instance, clever marketing might affect the system; customers might be convinced that the item is worth buying even at the price P_1. In other words, most interesting systems are complex and difficult to model. Nevertheless, it is often worth trying because the benefit of getting them right can be enormous.

The steps in the construction of a mathematical model (following Giodano, Weir et al. 1997) are:

1. Identify the problem.

2. Make assumptions
   a. Identify and classify the variables (for example, price and quantity).
   b. Determine the inter-relationships between the variables and any submodels (for example, in Figure 5.1 the variables are assumed to be in a linear relationship).
3. Solve or interpret the model. That implies working through the equations and inequalities that the model has generated.
4. Verify the model.
   a. Does it address the problem? Does it answer the questions you wanted to ask or have you strayed from that focus in constructing it?
   b. Does it make common sense?
   c. Test it with real-world data.
5. Implement the model.
6. Maintain the model.

Models are concerned with change. Some changes take place over discrete time periods (for instance, a company might review its prices annually). In that case, the model is embodied in difference equations and is said to be *discrete*. Alternatively, if changes occur all the time (such as population changes), the model is *continuous* and the equations are differential.

The above discussion assumes that it is possible to observe the system being modelled and to compare the model with the reality. However, there are many instances where this is not possible. It may be too inconvenient or dangerous or expensive to conduct experiments on the real system. It may also be that this is not possible because the system does not (yet) exist; it might be a planned telephone network. In such cases a *simulated* model might be appropriate. This approach is not new. For instance, the designers of a ship might have built a physical, scaled model which they would drag through a water tank and take measurements on. The difference now is that the model need not be a physical one, but a computer-based one. Modern new aircraft have been quite thoroughly tested on computers long before a single rivet has been punched through metal.

A number of programming languages have been developed specifically for the implementation of simulations. *Simula* (Birtwistle 1979) is probably the best-known one, but for reasons other than its usefulness in simulation – rather its introduction of the concept of object orientation.

Further reading

Giodano, F. R., Weir, M. D. and Fox, W. P. (1997) *A First Course in Mathematical Modeling*, Pacific Grove, CA: Brooks/Cole.

5.4   DEVELOPING TECHNOLOGIES

All (computer) technologies are in a constant state of development. The topics under this heading thus naturally represent a moving target; some novel topics that seem interesting now may turn out to be dead-ends, and there will be others that we cannot even conceive of at the moment. A useful starting point is the UK Computing Research Committee's Grand Challenges Exercise.[3] This is an attempt to identify the most useful and fruitful areas for future research, and it has identified the following topics.

In Vivo – In Silico

Computer systems continue to become increasingly complex. Much of Computer Science is concerned with managing that complexity. However, all around us are other systems of great complexity which have been around for millions of years: biological systems. Are there lessons we can learn from the way that Biology works that can be applied to computer-based systems?

There are different ways that this can be approached. One is to build computer simulations of life forms. On the one hand, the accuracy and success of simulations will tell us more about how biological systems work (or do not work), while on the other, they may give us new insights into ways of managing computer systems. We might create computers which evolve and improve, which are capable of repairing themselves, just as animals can recover from injuries.

Science for the global ubiquitous computer

At one time there was the computer in a room to which one or two people were allowed access at any time. Then the computer moved onto the desktop and had a single, dedicated user. Then the desktop machines were all linked together, via the Internet. This made vast resources available, but only as long as the users were in their offices. Then came wireless networks and computers that fitted in the pocket. So now anyone can have large amounts of computing power and almost infinite amounts of information available almost anywhere in the world. How do we ensure that we can manage all these resources and get the most from them? A whole new discipline is required to enable this to happen.

Memories for life

Advances in technology make it possible to store increasing amounts of information. One particular use that is being made of this facility is the storing of personal information: (**digital**) photographs, on-line biographies, audio recordings and the like. It has been calculated (Dix 2002) that within a lifetime it will be possible to store the entire life experience of an individual on a device the size of a piece of dust. How do we facilitate this, and what use do we make of it if we do?

Human memory is notoriously fallible. There are psychologists, though, who suggest we do store every experience we have – we are just not able to recover all of them. Would the same be true of a mechanised life memory? According to the Grand Challenges website:

> The challenge is to develop detailed models of an individual's abilities, skills, and preference by analysing his or her digital memories and to use these models to optimise computer systems for individuals. For example, a short-term challenge could be to develop a model of a user's literacy level by analysing examples of what he or she reads and writes, and linguistically simplify webpages based on this model; this would help the 20 per cent of the UK population with poor literacy. A longer-term challenge might be presenting a story extracted from memories in different modalities according to ability and preference; for example, as an oral narrative in the user's native language, or as a purely visual virtual reality reconstruction for people such as aphasics who have problems understanding language.

Scalable ubiquitous computing systems

The need for a new 'science' to manage ubiquitous computing has already been mentioned but other specific consequences of this development also need to be addressed. These include:

- Context awareness
- Trust, security and privacy
- Seamless communication
- Low powered devices
- Self configuration
- Information overload
- Information provenance
- Support tools
- Human factors
- Social issues
- Business models

The architecture of brain and mind

The most powerful and complex computer in the world is the human brain. There is no doubt that brains and computers work in very different ways. Might we make our artificial computers more powerful by making them operate more like brains?

Dependable systems evolution

Society's dependence on computing systems is increasing, and the consequences of their failures are at best inconvenient; in certain application areas, they may also lead to loss of financial resources, and even loss of human life. There are many challenges to realising and maintaining dependable systems. One of them is that once you have installed a system that is dependable, you want to rely on it for as long as possible; you do not want to change it too much in case you disturb its stability. That can mean that such systems remain based on obsolete technology. The challenge thus is not only to build systems that are reliable, but to build ones that retain their reliability as they naturally evolve.

Journeys in nonclassical computation

Currently they might include topics listed below; in time you might expect these topics to move into the list of mainstream topics in this section – and new ones to be identified.

Quantum computing
As discussed elsewhere in this book, computers work by manipulating electrical currents. Those currents obey the classical laws of Physics. That is to say, they can be described by equations such as Ohm's Law. In principle, though, it is possible to build computers that operate at the quantum level. At this level, the classical laws become special cases, and some effects that are very

unintuitive become apparent. As yet no one has been able to build a quantum computer, but theoretical studies show that if (or when) we can, then we will have an enormous increase in computing power. Most of what is said in Section 6.3: Data Structures and Algorithms, regarding the limitations of computers, would no longer apply; efficient (quantum) **algorithms** would become available to solve problems which previously only had solutions of exponential complexity.

Bioinformatics
A new approach to biology is based on the manipulation of vast amounts of data. The Human Genome Project is the best known example of this. It is only the availability of computing power that has made it possible to handle the data describing all of the millions of bases in DNA. There is great potential for biological and medical advances such as the design of drug molecules by computer.

Evolutionary computing
This is another link between biology and computing. Genes evolve through mutation and 'survival of the fittest'. The same idea can be applied to the generation of algorithms. Problems (which cannot be solved by simple algorithms) may be approached in stages by several algorithms that tackle them step-by-step. At each step, the result is evaluated and the algorithms that seem to be making best progress towards the solution are retained and mutated, while the less successful ones wither away.

Further reading

By their very nature, these topics cannot make their way into textbooks – yet. The best starting place, therefore, is probably the UK Computing Research Committee's Grand Challenges Exercise's website: http://www.nesc.ac.uk/esi/events/Grand_Challenges/

5.5   WEB-BASED COMPUTING

The Internet has been in existence since the 1970s, but much of what is done over the Internet only became widespread with the advent of the World Wide Web – the Web. Yet, the Web is not a technology, but an interface between the user and the network technology. Central to the Web is the *browser*, which is the implementation of that interface. The concept of hyperlinking means that the user can access all kinds of resources on the Internet (pictures, sounds, movies, other webpages and so on) by simply clicking on links.

The browser is often equated to the *client* because the Web is based on a *client-server* model. The server is essentially a computer on the network which contains one or more websites. The client (or browser) connects to the server and requests it to send materials – pages with their embedded content. Such forms of communication are, of course, not *ad hoc*; there have to be rules about how they take place, agreements as to the form and content of these communications. Such rules are known as *protocols* and web-based computing includes the study of the protocols used on the web.

A browser displays a webpage with formatting (large, bold text for headings and so on). Most such pages are created as simple text with formatting information embedded. The commonest format is known as HTML or *hypertext markup language*. Formatting commands are text enclosed in diamond brackets. For example, Figure 5.2 shows the 'raw' HTML for an extract of a simple webpage.

```
<h1>Sample Web Page</h1>
<h2>Introduction to HTML</h2>
<p>This is the start of a paragraph.</p>
<p>A picture of the file 'pic.jpg' would be displayed
here:</p>
<p><img src="pic.jpg"></p>
<p>and clicking on the word <a href="file2.html">link</a>
will cause the file 'file2.html' to be loaded.</p>
```

Figure 5.2 HTML code of a webpage that would appear as in Figure 5.3 when viewed in a browser.

You will see that most of the markup commands occur in opening and closing pairs. For instance, <h1> marks the start of a level 1 heading, which is closed with the corresponding </h1>. The other markup commands used in this example are explained in Table 5.1.

Markup	Explanation
h1	A level-1 (i.e. top-most) level heading.
h2	Second-level heading (headings can go down as far as level 6).
p	Paragraph
img	Image: marks the position of a picture on the page. The attribute 'src' must be given which specifies the name of the file containing the picture. There is no closing tag (i.e. there is no </img>).
a	Anchor: this marks an active piece of text, that will link to another file, such that clicking on the text will cause the other file to be displayed in the browser. The name of the file is specified through the attribute 'href'.

Table 5.1 HTML markup 'tags' used in Figure 5.2. Notice that most tags appear in pairs, so that (for instance) the beginning of a level-1 heading is marked by <h1> and terminated by </h1>

It depends on the browser how each of the marked-up elements in the HTML will be rendered. Figure 5.3 shows how

---

Sample Webpage

Introduction to HTML

This is the start of a paragraph.

A picture in the file 'pic.jpg' would be displayed here:

and clicking on the word <u>link</u> will cause the file 'file2.html' to be loaded.

Figure 5.3 How the HTML page shown in Figure 5.2 might appear in a web browser, such as Internet Explorer or Firefox.

this example might appear, but there is nothing to say, for instance, that the Level 1 heading must be that size or typeface.

A module on Web-based Computing would probably include an introduction to HTML. An HTML page such as that in Figure 5.2 can be created in a simple text editor, but there are also tools available to enable the creator to build and see the webpage as it might appear in a browser. Such tools are often referred to as **wysiwyg**, or 'what you see is what you get'.

Of course, not all webpages are simple hypertext, and they can be more interactive. There are a growing range of technologies available which are generally more akin to programming languages, and some of them might be introduced in a module on Web-based Computing.

As explained above, the Web is essentially an interface between people and the Internet. That does not necessarily mean that it is a good-quality interface. In other words, one aspect of Web-based Computing is human-computer interaction (HCI). While that is a topic that might be covered in a separate module, there are aspects that are particularly relevant in a web context. For instance, the strengths and limitations of HTML can affect the effectiveness of a webpage. Furthermore, there may be a particular kind of relationship between the owner of the website and the person visiting it. As might be covered in the topic of E-commerce, the website may be intended to sell goods to the visitor, but in order to do so a relationship of *trust* must be build between the two. Getting the design of the site and its interface right are important components in building up a relationship of trust.

Earning well-found trust also depends on making the website secure. Payment for goods is usually based on customers entering details of their credit card. Anyone who has those details can make charges against that card. It is therefore vital to ensure that the details are sent and are readable only by people whom the customer trusts. There are technologies to facilitate this, mostly based on encryption. That is to say, that when users enter their credit card details, it is not the plain text of that information that gets transmitted across the Internet, but an encrypted or 'scrambled' version, which can be decrypted only at the vendor's server. If anyone should intercept the data while it is en route across the Internet (and it is very easy to do this), it will be meaningless to the snooper.

The web is very much about sharing. However, sharing can be taken too far. Most material is owned by someone, who owns its copyright. That is to say that other people should not make copies of material without the permission of the owner. These rules are widely flouted on the web, though. The most common example is music, with services available through which people can download copies of songs to their PC and/or music player without any payment being made to the owner of the copyright of the material. New business models and new technologies are being created to find a compromise that will allow easier sharing of material without causing irreparable harm to the industries that create the material.

Further reading

Knuckles, C. D. and Yuen, D. S. (2005) *Web Applications: Concepts and Real World Design*, Hoboken: Wiley.

NOTES

1. It is probably only a matter of time before the hyphen is dropped from this term and we see *ecommerce*, just as *e-mail* is now *email*.
2. *Spam* derives its name from a sketch in the 1970s television comedy *Monty Python's Flying Circus*. The sketch was based on a café which insisted on serving the Spam brand of luncheon meat several times over with every dish. In the background of the café was a tribe of Vikings (it was a comedy) who repeatedly chanted, 'Spam, Spam, Spam,

Spam . . .' Presumably someone going down his or her list of arrived mail, seeing how much of it was unsolicited adverts, was reminded of this pointless repetition and coined the term.

3. http://www.nesc.ac.uk/esi/events/Grand_Challenges/

6 THEORY

6.1 ARTIFICIAL INTELLIGENCE

Artificial Intelligence (AI) is a discipline with two strands. One of them is related to psychology, and amounts to using the information-processing facilities of computers to study the nature of intelligence. The term 'cognitive' is often applied to this approach (Cognitive Science, Cognitive Modelling, and so on). We can build models and simulations of behaviour which is (or appears to be) intelligent and thereby learn about the way humans and other animals might achieve similar results. For instance, we can connect a video camera to the computer and then attempt to perform visual processing, so that items in the scene (for example, people or cars) are identified and classified. By getting a computer to perform some of these analyses, we might be able to find out something about how people achieve some of the same results.

The other side of AI is to try to make computers perform tasks with intelligence because it may be useful to have a machine to do this. In this context, the definition of what constitutes AI is a shifting one: it amounts to whatever is currently considered difficult computationally. For instance, programming language compilation is now considered a (relatively straightforward) topic in its own right (see Section 2.3: Compilers and Syntax-Directed Tools) but in the early days of high-level programming languages it was a task that had previously been performed only by people. Therefore, to get a computer program to carry it out was seen to be a manifestation of (artificial) intelligence. More recently, a scanner that you buy for less than £50 for your home computer will include optical character recognition (OCR) **software** that can 'read' printed text, yet it is not so long ago that OCR was thought to be a difficult AI problem.

So, AI tends to be applied to problems in which there is a lot of data and/or data that is uncertain. Some problems are so complex that it can be proved that no **algorithm** exists that will solve them – or in some cases at least not in a reasonable time (see Section 6.3: Data Structures and Algorithms, and Section 6.4: Theoretical Computing). These are one type of problem that AI can be applied to. A lot of effort goes into finding alternative efficient ways of solving such problems. These are often based on the idea of *heuristics*, which are really just rules of thumb. In other words, instead of using a straightforward – but inefficient – method, we try to find clever approaches. Yet AI is a discipline that has been affected like most others by the continued validity of **Moore's Law**. Whereas a few years ago, simple solutions were often impractical because of the time they would take, modern computers can run methods that do involve a lot of simple **brute-force** computing, but do so in a reasonable time.

Like any other discipline, Computer Science is subject to the vagaries of fashion. Within the discipline, AI is probably the most susceptible to fashion. The subject itself has gone though peaks and troughs of credibility over time. The idea of computers being able to perform novel and challenging tasks can be very exciting but, in this excitement, the advocates can over-sell their ideas. Then, if those ideas do not come to fruition, people may feel conned (not least those who funded the research), and the credibility of the subject falls. At present, though, the stock of AI is quite high: current achievements are impressive, and there have been highly-publicised feats such as computers winning chess contests against human grand masters.

The relationship between the two sides of AI inevitably means that it invites philosophical debate: what is intelligence, and how do we know if we find it – in people or in machines? Emotions are inevitably stirred up in this debate. For instance, many people are comforted by the idea that humans are the only entities with the power of thought and they may feel threatened by the suggestion that machines could be made to think. There are two major camps, characterised as *strong AI* and *weak AI*.

The strong AI position is that it is – or will be – possible to make computers think. By extension, advocates of this position might suggest that the human brain is nothing more than a complex computer, which is equally bound by the limitations on computability that machines are – although our limitations are less apparent because of the power and complexity of our brain 'computers'.

The weak AI position is that we are interested in producing behaviour in computers that *appears* to be intelligent. There are programs that can beat grand masters at chess, but no one would claim that the computer *thinks* about playing chess. This is a more utilitarian approach: if the computer can do a better job by appearing to be intelligent, then it is irrelevant that it is not thinking.

AI is a dynamic discipline. The topics studied change as what was once considered hard becomes everyday and as new, more difficult challenges are tackled. The success of the subject rises and falls too, depending on what is promised and what is achieved. New approaches emerge and old ones are re-named. At the time of writing, there is great interest in the topic of *agents* (intelligent agents, rational agents . . .) Already agents are becoming almost commonplace (they are found in the logic of washing machines, for instance), so perhaps research will soon move on to new approaches – or back to old approaches with new names.

Further reading

Cawsey, A. (1997) *The Essence of Artificial Intelligence*, London: Prentice Hall.

Russell, S. J. and Norvig, P. (2003) *Artificial Intelligence: A Modern Approach*, Upper Saddle River, NJ: Prentice Hall.

6.2   INTELLIGENT INFORMATION SYSTEMS TECHNOLOGIES

One of the effects of the widespread adoption of information technology is that companies accumulate large amounts of data. Such data has to be of value, and yet the sheer amount of it can make it extremely difficult to find what is relevant. As Peter Fabris has eloquently described it:

> Have you ever seen one of those posters that at first glance looks like a jumble of colored dots? Stare at it, and a three-dimensional picture will jump out from the pointillistic background. Now, think of those dots as the bits of information about your customers contained in your company's databases. If you look at the dots of information in the right light and at the right angle, they will reveal patterns that yield insight into customer behavior.
>
> (Fabris 1998, original spelling).

The technique he goes on to describe is that of *data mining*. This amounts to applying statistical analyses to the collected data in order to shed 'the right light and at the right angle' on it so that the owner can see the useful information contained within it.

All the data that a company holds is likely to be held in a variety of forms by a number of departments. They will have different databases, based on different technologies, tailored to meet their own needs, but not necessarily compatible with the needs of other departments. They will have been built up at different times and are based on different technologies – they are often referred to as *legacy systems*. This complicates the task of the data miner but it can be addressed with the use of *data warehousing*.

Data warehousing emphasises the capture of data from diverse sources for useful analysis and access, but it does not generally start from the point of view of the end-user or knowledge-worker who may need access to specialised, sometimes local databases. The latter idea is known as the *data mart*. To use a simple analogy, it is like finding the proverbial needle in the haystack. In this case, the needle is that single piece of intelligence the business needs and the haystack is the large data warehouse the company has built up over a long period of time.

Information in this sense is less precise than data. It is closer to the way humans think about things than the way computers operate. Correspondingly, a number of techniques and methods have been developed that enable computers to operate in a less precise but humanly meaningful way. For instance, it is often stated that computers are very logical in operation, but conventional logic is sometimes too precise to cope with inexact inputs and the contradictions of human life. Therefore, a special type of logic has been developed: its name – *fuzzy* logic – reflects the fact that it operates in a less precise manner than traditional logics.

Another technique is that of *case-based reasoning*. One definition of an expert might be someone who has 'seen it all before'. In other words, when a problem arises, an expert will (subconsciously) review his or her memory of similar problems in the past. In this way, the expert may recall the solution that worked before and see if it can be applied or adapted to the current challenge. Those previous instances are *cases*, and case-based reasoning relies on saving representations of cases on computers, matching them to subsequent similar cases and presenting a proposed solution.

The success of any enterprise depends on the making of decisions. The quality of a decision can depend on information. Information technology means that a great deal of information is available, but there is so much that it can overwhelm the decision-maker. It is thus critical to be able to design systems that provide an accessible, usable interface between the information and the people who would use it.

Further reading

Callan, R. (1998) *The Essence of Neural Networks*, London: Prentice Hall.

Inmon, W. H. (1996) *Building the Data Warehouse*, New York: Wiley.

6.3   DATA STRUCTURES AND ALGORITHMS

An **algorithm** is a set of instructions that can be followed precisely to achieve some objective. That description sounds very much like a computer program and, indeed, algorithms are a way of studying programs in a way that is independent of implementation details, such as the programming language or computer **hardware**.

A typical example (studied in any module on algorithms) is *sorting*. That is to say, there are many applications in which we need information in order (listing names in a phone directory, checking your lottery numbers, and so on). There is more than one way – or algorithm – that can be used to sort items.

We are interested in properties of algorithms, usually how much time and how much memory a given algorithm will need. Algorithms are generally measured in terms of the numbers of items of input, usually referred to as n. For example, the size of the input to a sorting algorithm is the number of items to be sorted. An *efficient* algorithm is one that takes time proportional to a polynomial function of n, such as n^2. Some problems have less efficient solutions. For instance, they may take time that is exponential, such as 2^n. At first glance, that might not seem to be very inefficient (especially given the ever-growing power of computers) but in practice even the most powerful computer cannot run such algorithms in realistic time. Such problems are impractical to solve algorithmically and become objects of theoretical interest (see Section 6.4: Theoretical Computing) or are solved using the techniques of artificial intelligence (see Section 6.1: Artificial Intelligence).

Within the set of efficient algorithms, it is still useful to be able to compare different solutions to the same problem. Most sorting methods take time proportional to n^2, but there are some that approach n in some circumstances (such as when their input is already nearly sorted). Thus, it may be better to choose the algorithm that can run in time n rather than one that always takes n^2.

Time is not always the sole matter for concern. Often it is possible to trade time against memory space, and that will depend on the *data structures* used to represent the information. For instance, to represent postal addresses you might simply store the words that make up each address. Alternatively, you might group together all the addresses in one county, and then all the addresses in each town and so on. Each representation would take a certain amount of memory and would make some operations easier than others. Storing all the text of the addresses would take up a lot of storage because the name of the county would be stored once for every address in that county, but it would be easy to print out each address. Using the second data structure, it would be easy to find all the addresses in one town. The trade-off is usually that saving space in the data structure means that it takes longer to process it.

There is another set of problems for which we can find polynomial-time algorithms, but only by using a trick. The trick is to assume that you could have a computer that would be able to correctly 'guess' the solution. This ability to randomly choose the right solution is known as **non-determinism**. Of course, no non-deterministic computers really exist: they are a useful basis for some thought experiments. It would seem that problems that could only be solved in polynomial time if there were non-deterministic machines must be harder to solve than those that can be solved on a real, deterministic machine. In other words, if the set of problems with non-deterministic polynomial solutions is NP and the set of problems with deterministic polynomial solutions is P, then surely those two sets are different, so that NP ≠ P. It seems unlikely that NP = P, and yet no one has been able to prove it either way. This remains an open research question. There are great prizes awaiting the computer scientist who can prove that NP = P, or not.

Further reading

Sedgewick, R. (1988) *Algorithms*, Reading, MA: Addison-
Wesley.

6.4   THEORETICAL COMPUTING

We have discussed above whether the subject of this book is
really a science. If it is a science, then it should have a theoretical basis. Physics, for instance, has matured as a science
with definite theoretical and practical aspects. Sometimes
theory has to be developed to explain observed experimental
results and sometimes the consequences of a theory suggest
new areas in which the practitioners should look for confir-
mation of the theory. Again, though, Computer Science does
not have this characteristic. Instead, there are a set of mathematical results, tools and techniques, some of which have
profound implications for what can or cannot be done on computers, and some of which are useful tools for the Computer
Scientist. Some of those topics might be taught in a separate
module, but often they would be included within some of
the topics discussed elsewhere (for example, Section 6.3: Data
Structures and Algorithms, and Section 2.3: Compilers and
Syntax-Directed Tools).

One of the most important tools for theory is the Turing
Machine. This is not a real machine but a theoretical one, a
basis for thought experiments. It was 'invented' by mathematician Alan Turing (who was surely the first theoretical
Computer Scientist) even before any real computers existed
(Turing 1936). He set out what he saw as the most elementary
possible computer. It includes a tape containing symbols
(zeros, ones and blanks). The symbols can be written on the
tape by the machine, and also read by it. The machine can be
in any one of a number of *states*. The machine has a program
which is just a set of instructions of the form:

If in state *S*, with symbol *B* on the tape then perform action *A*.

Actions allowed include:

- move the tape one place to the left
- move the tape one place to the right
- read the symbol in the current cell
- write a 1 on the tape
- write a 0 on the tape.

This may seem a very simple, crude machine, and indeed it is. Be aware, also, that it is not a very *convenient* approach to computing – it is merely *possible* to do it this way. However, one of the most important results in theoretical Computer Science, known as the *Church-Turing Thesis*, is that no computer that could ever be built could be more powerful (in terms of what it can and cannot compute) than a Turing Machine. Put another way, it means that all computers are equally limited. Your home PC may not be as expensive as the supercomputer that the Met Office uses to generate the weather forecast, but there is nothing that their computer can do that yours cannot, or that could not be run on a Turing Machine. Of course, the reason the Met Office spends so much money on their computers is that they need the results fast. There is no point in generating a forecast for the next day that take twenty-four hours to compute – in that case, it is no longer a forecast! Instead of taking twenty-four hours on your PC, it can be done on their computer in, say, one hour – but there is no reason in principle why the same computation could not be completed on your PC.

As computers are being used increasingly in different roles, it is easy to gain the impression that there is no limit to what they can do. Indeed, some people have publicly suggested such.[1] This is fallacious. There are limits to what can be done by computer. It is important to be able to reason about those limits, and that is one of the roles of theoretical Computer Science. In particular, there are problems that do not have computable solutions, and it can be proved that it is not possible to write a computer program to solve them. These problems are said to be *undecidable* or *noncomputable*.

It is possible to prove mathematically that such problems cannot be solved on any computer – no matter how big, expensive or powerful it might be. The first-known such problem is called the *Halting Problem*. Like many difficult problems, it sounds deceptively easy. Anyone who has used a computer will at some time have experienced a crash. Sometimes the crash is apparent because an error message appears on the screen, but sometimes the program 'hangs'. It appears to be doing nothing and is unresponsive to any inputs you try to give it. In fact, it is doing nothing very fast – it is stuck in a loop. It would be useful if it were possible to write a program that can test other programs and say whether it is the case that the tested program will always halt – never become stuck in a loop – for all the possible inputs to that program.

In fact, it is possible to prove that it is logically impossible to write such a checking program. This was a result first proved by Alan Turing – before computers existed – and described in terms of his imaginary Turing Machine. There is a whole set of problems which are undecidable. Many of them are equivalent to the Halting Problem. If one can demonstrate that a given problem is equivalent to the Halting Problem, then one has proved that it is undecidable. Clearly, before you set about devoting time and money to the development of a new piece of **software**, it is as well to check whether it may be impossible to write it at all.

There is nothing we can do about undecidable problems. It makes little sense even to try to approximate answers to them. You might tackle an instance of the Halting Problem by running the program under test with one million sets of input data. If it halts for all of them, you might think you had nearly solved the Halting Problem, but you will never know if the program would have looped on the next set of data.

Undecidable problems are not the only ones that cause difficulties. We have seen in Section 6.3: Data Structures and Algorithms, that we are concerned with finding efficient solutions to problems where an efficient solution is one that takes no more than polynomial time and space, in terms of the number of inputs. We can show mathematically that a solution (algorithm) that takes n inputs will take time proportional to, say, n^2 to run. Some problems have solutions that we can find, but that are more complex than that. An algorithm with time complexity n^3 is clearly going to take longer than the n^2 one, but we can probably live with that; but what if the complexity is (say) 2^n? That function grows much more quickly (see Figure 6.1). If $n = 64$, then $2^n = 1.84 \times 10^{19}$, or 18,400,000,000,000,000,000.[2] Suppose that the computer could execute one million of the instructions required to solve this problem in a second, then it would take 1.84×10^{13} seconds, which is 583,000 years.

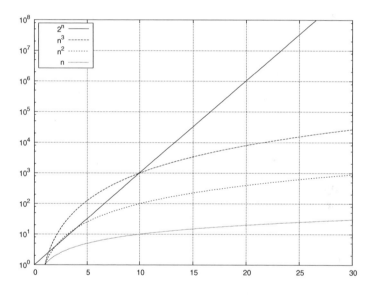

Figure 6.1 Comparison between different growth rates of complexity measures. Note that the vertical axis is a logarithmic scale, which shows that the 2^n curve grows very much more quickly than the others.

There are some important facts to bear in mind here. Many apparently simple problems have solutions with complexity of this form (known as *exponential* complexity), and here we have considered a value of n as small as 10, and assumed that we have a computer that can execute a million instructions per second. Even then we are talking of timescales that are clearly impractical. Double n to 128, and $2^n = 3.40 \times 10^{38}$, and the corresponding execution time becomes 1.08×10^{25} years (10,800,000,000,000,000,000,000,000 years). To put that in context, universe is thought to be 13.70×10^9 years old.[3]

When we are talking about such large numbers, it is evident that computing power is not the answer. Recall **Moore's Law**, which states that computing power doubles every eighteen months. That means that a weather forecast that can be generated in one hour today would have taken sixteen hours on the computers available six years ago. In other words, it is practical for the meteorologists to use more complicated weather models fed with larger amounts of data to generate a more accurate forecast than was feasible six years ago (when sixteen hours to produce a twenty-four-hour forecast would not have been very useful). So, it is worthwhile for the Met Office to buy the most powerful computers they can afford, but they will still be unable to run inefficient algorithms.

Let us return to the problem with a 2^n complexity solution. In eighteen months' time, we should be able to execute double the number of instructions per second – in other words two million – so we can divide all the numbers above by two. So, the more powerful computer will make it more practical to run complex but efficient programs, but makes no difference to the inefficient ones.

An important contribution that the theorist can make is to identify that a problem only has algorithms that are inefficient (that is, they are worse than polynomial). This time, though, we might tell our programmer not even to bother starting, but perhaps to think about solving a simpler version of the problem. For instance, we might be content with a solution that works for most – but not all – inputs, or one which we know will give something close to the best possible (optimal) answer, but is not guaranteed to come up with the best one. In general, if we have got to this point – the problem is interesting and worth solving, even if it is not possible to derive a perfect solution – then we are moving towards the realms of

Artificial Intelligence, which tends to take a very different approach to its problems.

As you will know, becoming stuck in a loop is not the only way that a program can fail. It is frustrating for the average user that a computer crashes from time to time. On a larger scale, there are innumerable examples of large amounts of money having been lost when a program has gone wrong. People often wonder why this is so, why it seems impossible to write programs that are completely free of errors. The fundamental reason is that programs are very complicated. For some pieces of software, it is particularly important to be as sure as possible that it contains no errors. While it may be annoying for you that your word processor has crashed, lives may depend on the program controlling an airliner being free of errors. One approach to ensuring good, bug-free software is to build a mathematical representation of the program and then to prove its properties using the rules of logic – just as you might prove the correctness of a mathematical theorem.

As discussed in Section 2.5: Programming Fundamentals, the meaning of a computer program is referred to as its *semantics*. There are three main approaches to the expression of program semantics: operational, denotational and axiomatic. According to Nielson and Nielson (1999), the differences between these approaches are:

Operational semantics The meaning of a construct is specified by the computation it induces when it is executed on a machine. In particular, it is of interest *how* the effect of a computation is produced.

Denotational semantics Meanings are modelled by mathematical objects that represent the effect of executing the constructs. Thus, *only* the effect is of interest, not how it is obtained.

Axiomatic semantics Specific properties of the effect of executing the constructs are expressed as *assertions*. Thus, there may be aspects of the execution that are ignored.

What all these approaches have in common is that they allow people to express properties of programs in mathematical notations – and mathematics has the power to make assertions and proofs about properties in general. So, for instance, it can be possible to prove that a particular program will always behave in a particular way, regardless of the language it is written in and the **hardware** on which it runs. In some senses, it would be ideal to do this for all software, to ensure that it is correct and bug-free.

This approach is referred to as *formal methods*. The problem with it is that it is very difficult to do, which is why it tends to be applied only to important (for example, safety-critical) software. The amount of effort (and money) needed to write and prove the mathematical specification is likely to be many times operator than that required to simply write the program in the conventional (but bug-prone) way. However, there are tools to help. Mathematical proofs do not have to be carried out with pencil and paper because there are theorem-proving programs that can be used to help perform the proof (though they are not sufficiently powerful to do the whole proof automatically).[4]

It should be evident from this discussion that Theoretical Computer Science is very much based on Mathematics. The emphasis on Maths will vary between different Computer Science programmes, but many will include modules on Maths. You should be aware that the kind of Maths covered is likely to be quite different from that which you have studied at school and in the sixth form. If you have enjoyed Maths so far, that is a good sign, but be aware that some of the topics you will cover at university are likely to be new and somewhat different.

Whereas much of the Maths you have studied at school may have been concerned with *continuous* quantities (smooth curves on graphs, from which you can calculate gradients and the like), computers often deal with *discrete* entities, and there is a branch of Mathematics which deals with that. So, for instance, the kinds of proofs in formal methods are logical ones, where you deal with the simple *boolean* values, *true* and *false*.

Further reading

Currie, E. (1999) *The Essence of Z*, London: Prentice Hall.

Dean, N. (1996) *The Essence of Discrete Mathematics*, London: Prentice Hall.

Kelly, J. (1996) *The Essence of Logic*, London: Prentice Hall.

Nielson, H. R. and Nielson, F. (1999) *Semantics with Applications: A Formal Introduction*, Chichester: John Wiley.

Nielson and Nielson (1999) deals specifically with the topic of semantics. It has the advantage that it can be downloaded freely from the Internet, subject to certain conditions (http://www.daimi.au.dk/~bra8130/Wiley_book/wiley.html).

NOTES

1. Harel, D. (2000) *Computers Ltd: What They Really Can't Do*, Oxford: Oxford University Press, includes the following quote from *Time* magazine, 'Put the right kind of software into a computer, and it will do whatever you want it to. There may be limits on what you can do with the machines themselves, but there are no limits to what you can do with software.'

2. Strictly speaking, we should write $2^n \approx 1.84 \times 10^{19}$. In other words, that number is an approximation to the value of 2^{64} and there could be many more digits after the decimal point. The fact that we can only represent numbers approximately – particularly very large and very small ones – is something that Computer Scientists struggle with all the time, and is touched on in Section 3.1: Computer Architecture.

3. http://map.gsfc.nasa.gov/m_uni/uni_101age.html

4. Of course, there is a worryingly circular conundrum. If I use a theorem-proving program to prove that a piece of software is 'correct' (that is, bug-free), how do I know that the theorem prover itself does not contain bugs that will mask bugs in the program I am proving?

7 SYSTEMS

7.1 COMPUTER-BASED SYSTEMS

A system is a (usually complex) combination of elements which interact with each other in different ways. A fundamental requirement in many areas of business, economics, biology, engineering and politics is to be able to understand systems and to control them. There is a large body of work on Systems Analysis, and increasingly computers are part of nearly all artificial systems. It is therefore vital to study and understand systems – as far as possible.

There are many different kinds of systems; some of them are described below.

Embedded systems

These are systems that include computer technology, but the computer is largely invisible ('embedded'). For that very reason, people are often unaware that there is a computer involved. The modern car contains a large number of computer processors – managing the engine, the suspension, the anti-lock braking system and so – on, and yet few people would describe a car as a computer system.

Real-time systems

These are systems linked to events in the outside world. Those events take place according to imposed time constraints and it is necessary that the computer component of the system keeps up with the timing of those events in the 'real world'.

An important example is air traffic control. Given the course and speed of an aircraft, there is a certain time before it will reach a certain point in the sky at which it will have to change course. The computer system that monitors and controls this situation must calculate the necessary change before the aircraft reaches that point; if the computer is too slow, then the aircraft will pass the point before the new course has been calculated. If the reason for the course change is to avoid a collision with another aircraft, then being late with the calculation could be disastrous.

Safety-critical systems

The air traffic control system is an example of a safety-critical system. Clearly, special care has to be taken in the design and implementation of such systems. Engineering approaches may be applied more carefully to such projects. Another approach is the application of formal methods (see Section 6.4: Theoretical Computing). These may be used to ensure or prove that the most critical parts of the system are absolutely bug-free. Risk analysis can be another important component of the development of a safety-critical system. Given that in most cases it is not possible to assert that any system is 100 per cent reliable, the question is what is the probability of failure – and then what is the likely cost of such a failure.

Distributed systems

The components of a computer system (processors, memory, and so on) need not all be contained in a single box; they may be distributed in various ways. Such systems are so common and important that they are often studied separately – see Section 4.5: Distributed Computer Systems.

Client-server systems

One particular form of distributed system is based on two kinds of computers: *clients* and *servers*. A server generally has a rich set of resources (memory, discs, processing power) available. A client has fewer resources, but can request a share of resources from servers. The Web is the most common example of a client-server system. Your PC, running a browser (Internet Explorer, Netscape or whatever) is a client, but it requests pages – with images and multimedia and all the rest – from web servers around the Internet.

Many of the topics relevant to computer-based systems might also be covered in the topic of Systems Analysis and Design and are discussed further in that section (7.3).

Further reading

Berger, A. (2001) *Embedded Systems Design: A Step-by-step Guide*, New York: Osborne McGraw-Hill.
Burns, A. and Wellings, A. J. (2001) *Real-Time Systems and Programming Languages*, Reading, MA: Addison-Wesley.

7.2   DATABASES

Computer Scientists often draw distinctions between *data*, *information* and *knowledge*. The three are closely linked, but the distinctions are important. As is often the case, though, there are no universally-agreed definitions of the terms, so you have to be careful in trying to adhere to any particular definition. However, the understanding of this section will be easier if we agree on some definitions of these words.

'Information' is usually used as the most general description of facts. Data is the most elementary kind of information. Generally, data is that which can be represented by numbers. A database is a computer store of data. A patient record database might keep records of measurements taken from patients, such as temperature, blood pressure, blood sample analyses and the like. These will be recorded as exact numbers.

Confusingly, databases do not only store data. Another part of the database might include free-text-format *information*, information not representable as numbers, such as the patient's mood, demeanour and other non-specific symptoms. A doctor might use all that is stored about a patient, along with observations and his or her own experience, to diagnose the patient's condition. In so doing, the doctor is making use of *knowledge*. As the name implies, a database stores data, which will be described in more detail in this section. Information systems are described in Section 7.4.

As so often is the case in Computer Science (and other disciplines), the distinctions between data, information and knowledge are not always as clear-cut as might be convenient. Observations on the patient's mood and demeanour consist of text which comes under the category of information, but which is nevertheless stored in a computer. In other words, *information* is stored in the form of (**digital**) *data*. As we will see in this and other sections, a lot of the work that Computer Scientists do is about transformation between the categories of data, information and knowledge. So, for instance, many books (particularly historical ones) have been scanned into a form that can be stored and viewed on computer. Such *digitising* of books can be viewed as transforming information into data. On the other hand, *data mining* (see Section 6.2: Intelligent Information Systems Technologies) might be seen as turning data into information.

There are different ways of organising data in a database, but the most common is the *relational* model. A relational database consists of *tables* or *files*. Each table contains a set of *records*, and each record is made up of a set of *fields*. For instance, a database of patients might have a *names* table. A name is a record that consists of two fields containing the first name and the surname. Of course, it is quite likely that two people of the same name will appear in a database if it is of any size. This is likely to cause problems. Adding more information, such as middle names and/or date of birth, will make identical records less likely, but really one has to be certain that there is a unique way of identifying each individual in the database. That is why numbers are often generated and allocated to people. Every patient should have an NHS number and that might be used as the unique *primary key* in the patient databases.

In a real database more information will be required on each individual. Their residential address might also be stored. This might be in a separate *address* table. This will contain fields for the house number, street, town, county and postcode (see Figure 7.1).[1] Of course, most homes have more

Names table

First name	Surname
Liz	Windsor
Ken	Hyde
Horace	Windsor
Gurmit	Singh
Jack	Knox
Chris	Keough

Addresses table

No	Street	Town	County	Postcode
1	Lilac Av	York	N Yorks	YO1 1DD
42	High Street	Leeds	W Yorks	LS8 8XX
23	Blossom St	Leeds	W Yorks	LS1 1AA
10	Circus Close	Hull	Humberside	HU4 7ET
6	Race Road	Boston	N Yorks	YO6 2AB

Figure 7.1 The structure of a relational database. This one consists of two tables: *names* and *addresses*. Each table consists of a set of *records* and each record has a number of *fields*. There would be links between the fields of one table and the other, but there is not necessarily a one-to-one correspondence. For instance, Liz and Horace Windsor might be a married couple living at the same address, so that they would share a record in the addresses table.

than one resident. That means that each address need only be stored once in the address table and each of the residents in the names table can be linked to the same address record in the addresses table. Before building a database, the designer will generally work out the nature of the relationships between all of the records that are going to be stored in it by building an *entity relationship model*, which can be represented visually in a diagram. By drawing such entity models, the designer can ensure the database has a structure that matches the data to be stored and that will be efficient and safe. This is facilitated by formal manipulations of the model which will transform them into *normalised* forms.

A database can be viewed as sets of tables stored in a structured way in computer files. In order to access the information and manipulate it, you need a *database management system* (DBMS). This will allow the user to extract required information and to process it (for example, to extract the names and addresses of every person in the database living in a given town and present the information in alphabetical order of surnames). Different database managers will allow the user to interact in different ways. One might present the information in a manner that resembles a webpage, while another might allow them to pick records directly using their PC mouse, for instance. However the user interacts with the DBMS, the software will generate *queries* in a specialised query language.

A database is not merely a form of storage. To be useful, it must be possible to extract information from it. Given the precise form of data, it is possible to extract exactly the information the user requires. To do this, the user constructs a precise specification of the information required in a database query language. The most widely used query language is *Structured Query Language*, or SQL. For instance, the postcodes of all the patients living in York, recorded in the *addresses* table of a database, might be retrieved by the query:

```
SELECT Postcode FROM addresses WHERE Town =
"York"
```

This is just a simple example. Much more complex ones can be generated. For instance, the user might want to know the names of all of the patients treated by a given doctor between certain dates, and he or she could obtain this by composing a similar – but more complex – SQL query. Part of learning about databases might include crafting appropriate SQL queries by hand, but remember that in practice when using a real database, the database manager will generate the SQL for you.

Databases exist to store specific types of data. For instance, there has be a great increase in the collection and storage of information that is geographically related in Geographical

Information Systems, or GISs. For example, retailers can work out the optimal site for a new supermarket outlet by using a database containing information about the distribution of potential customers, existing supermarkets and so on.

Some information can be difficult to store and access in a structured way. Anything that comes under the label of 'multimedia' is an example. Video footage is very rich in information. Why any individual might want to access a given piece of video may be very different from why another wants to do it: one might be interested in the principle, foremost person in the film, another might be interested in the commentary, and another might want to see the scenery in the background. Storing such information in a way that all these three could find what they need – without the entire content having to be laboriously indexed and described in words by human operators – is a major requirement and the subject of continuing research.

The information stored in a database will often be highly critical for the owner's business. For instance, if invoices are not issued, they will not be paid and the company loses money. Similarly, an invoice sent to the wrong address may never be paid. Such errors can occur through operator error (for example, someone typing in the wrong address) but it is also vital that the DBMS does not generate any errors in the database. This comes under the term of *data integrity*. A great deal of effort goes into establishing and maintaining the integrity of the data.

Security is a related concern. Often, some or all of the information in a database is sensitive and/or confidential. For instance, the balance in an individual's or a company's bank account should not be accessible to anyone other than the account's owner or the bank. Beyond that obligation, there is legislation stating that *any* information that identifies an individual person must be treated as confidential. The responsibility for confidentiality on a database administrator is quite onerous and a great amount of effort must be expended on maintaining security. Evidently, this topic overlaps with the material that might be covered in modules on topics such as Security and Privacy.

Modern databases are not necessarily held on a single computer. Often, they must be accessed from multiple locations (many travel agents will use the same booking database), and therefore many of the topics covered under topics such as Distributed Computer Systems (see Section 4.5) are also relevant.

Further reading

Carter, J. (2003) *Database Design and Programming with Access SQL, Visual Basic and ASP*, London: McGraw-Hill.
Date, C. J. (2003) *An Introduction to Database Systems*, Reading, MA: Addison-Wesley.
Rolland, F. D. (1997) *The Essence of Databases*, London: Prentice Hall.

7.3   SYSTEMS ANALYSIS AND DESIGN

The effective handling and presentation of data and information depends on the use of computer systems. Any organisation is likely to have a number of such systems, of different types, including:

Transaction processing systems (TPS)
These handle routine business transactions, such as payroll, inventory and ordering.

Office automation systems
These include familiar tools such as word processors and spreadsheets.

Management information systems (MIS)
These are computerised information systems designed to support management, based on a database containing both data and models.

Decision support systems

These are management information systems that are specifically designed to support decision making.

Expert systems and artificial intelligence

When any problem is too complex to be tackled by systematic algorithmic approaches, artificial intelligence (AI) may be applied. Expert systems represent one particular branch of AI. The behaviour of human experts is captured and analysed and then embodied in rules which can then be called on to solve similar but novel problems.

Evidently, if any company is going to use all of this variety of systems (and there are others that might be added to the list), it should do so in a clean, integrated manner. It is the role of the Systems Analyst to ascertain the necessary structures, to design systems to support them and to oversee the implementation of them (the writing of the programs will be left to programmers, under the Systems Analyst's direction). Such a complex collection of inter-related systems could easily become chaotic. It is the Systems Analyst's job to prevent that and to bring order, and they must therefore have a structured approach to the problem. A number of such approaches have been devised, but one of the most common ones is the *waterfall model*, illustrated in Figure 2.8.

The role of the Systems Analyst is very broad. This is apparent in the waterfall model, which implies that the analyst must see the process through all the way from specification to maintenance. Techniques and methods have been developed for each of these stages. Some of these methods are quite formal (that is, mathematical) while others are less so. The analyst must deal with organisations of people and how they work and interact. Their analysis will lead them to design systems that fulfil the identified needs. Those designs will have to take account of the abilities and limitations of computers on which they will be implemented, so that the analyst should direct the use of appropriate software engineering processes and tools, as well as being aware of the appropriate human-computer interface requirements. In other words, the Systems Analyst requires a high level of knowledge of a number of the topics described in this section.

Further reading

Booch, G. (1994) *Object-Oriented Analysis and Design*, New York: Addison-Wesley.
Griffiths, G. (1998) *The Essence of Structured Systems Analysis Techniques*, London: Prentice Hall.
Kendall, J. E. and Kendall, K. E. (2002) *Systems Analysis and Design*, Upper Saddle River, NJ: Prentice Hall.

7.4 INFORMATION SYSTEMS

The systems listed above (Transaction processing systems, and so on) are all examples of information systems. While their study is an inherent component of Systems Analysis, they may be studied in their own right. Topics that might be included in such studies include the following:

- Information in organisational decision making.
- Integration of information systems with organisational strategy and development.
- Information systems design.
- Development, implementation and maintenance of information systems.
- Information and communications technology (**ICT**).
- Management of information systems and services.
- Organisational and social effects of ICT-based information systems.
- Economic benefits of ICT-based information systems.
- Personal information systems.

Clearly many of these topics overlap with other subjects described in this section.

Further reading

Ratzan, L. (2004) *Understanding Information Systems: What They Do and Why We Need Them*, Chicago: ALA Editions.

7.5 SECURITY AND PRIVACY

If 'Knowledge itself is power' and 'Power corrupts', then in the modern age information can lead to corruption. Thus, it is necessary to develop techniques and technologies that can maintain privacy and security to counter those who would misuse the technology.

Encryption is one aspect of this and it is an example of a fascinating problem, which is founded on some complex mathematical theories. Most 'secret codes' are based on encryption. That is, a message is transformed into an apparently meaningless code which can only be unscrambled back into the original text by someone who knows the encryption method that was used and also a secret key. Encryption is becoming increasingly important as increasing numbers of sensitive messages are transmitted over networks. At one time, someone could only spend money from my bank account if they got hold of my cheque book and my bank card. Now they might get all the information they need by intercepting a message that I send to an on-line retailer – unless that information is encrypted.

No encryption system is entirely uncrackable.[2] The level of security thus depends on the amount of time and effort the code breaker can put in. If the encrypted information is out-of-date by the time the encryption has been cracked, then it has been successful.

Further reading

Russell, D., Gangemi, G. T. and Gangemi, G. T., Sr (1991) *Computer Security Basics*, Sebastopol, CA: O'Reilly.

NOTES

1. Of course, not all addresses fit into that format; some houses have names not numbers, for instance. It is one of the challenges of the database designer to be able to cope with the diversity of the real world, which may not fit easily with the model of the world embodied in the database design.
2. There are some that are mathematically unbreakable, but even they can be compromised. For instance, the method known as the *one-time pad* is mathematically unbreakable, but it relies on the sender and receiver both having a copy of the codebook (the 'pad'). Hence, it can be broken if someone else obtains a copy of the pad.

PART II
Thinking of a Computer Science Degree

8 PREPARATION

8.1 COMPUTER SCIENCE, STUDIES, IT . . .?

Recall the list of topics in Chapter 1: The Elements of Computer Science. Any Computer Science programme that you investigate will include some subset of these topics; none will include them all. You need to look at the different sets of topics offered and pick those that you believe best suit you and your aspirations.

In this part of the book we look more at how Computer Science is taught within degree programmes and at how to apply to get on those programmes. The aim is to give you the best chance of ending up on a programme which suits you.

All university undergraduate admissions are handled by the Universities and Colleges Admissions Service (UCAS), more details of which you will find in Chapter 9: Applying to University.

One of the first problems you will have with choosing a programme is the names of them. UCAS lists over 2,800 programmes under the heading 'Computer Science'. With as large a number as that, it is difficult to be precise as to how many different names are used, but a rough calculation suggests there are over 500 different related programme titles. Most of them have titles such as 'Computer Science' and 'Computer Science with *something*'; the latter includes some that are quite esoteric and offered only at one institution, such as 'Business Information Systems and Sport & Exercise', 'Computer Science and Dance' and 'Computer Science with Nutrition' (one for the late-night hackers who live on Mars bars and delivered pizzas). The point is that if you were looking for a course on Physics or English, then those are the

titles you would look for and there would not be so many
variations – apart from the various 'Physics-with' combinations. However, if you want to study a subject related to computers, you will have to take a closer look at exactly what the
different programmes comprise.

'Computer Science' is not the only term to look for. There
are a number of related disciplines which might be closer to
that which you would like to study. These could include topics
such as Software Engineering, Informatics, Information
Engineering, Information Technology and Computer Systems.
(Notice also that Scottish universities seem to prefer the term
Computing Science.) Trawling through the 2,800 programmes
mentioned above might be sufficiently daunting that you are
reluctant to broaden the scope of your search even further. On
the other hand, by doing so, you may find that other programmes (that you had not even thought of before) are closer
to your interests.

We can now discuss some of the things that Computer
Science is – and is not.

Software Engineering
This is another term that is used and misused by different people.
It is often used as a posh term for programming, but what
most academics would suggest is that it is the application of engineering principles to the production of software. When a civil
engineer is commissioned to build a bridge, he or she does not
go straight out and order concrete and steel. First, the engineer
draws up plans and performs stress calculations to predict
how the design will work. By contrast (in the past, at least) a
programmer might start straight away with his or her raw
materials – a program editor and immediately start writing code.
And that is why programs often have bugs. The problem is that
the civil engineer has some basic principles (the laws of Physics)
and generations of techniques and practices which can be
used to ensure that the bridge will behave in the way predicted. Programmers have fewer such foundations, but the
Software Engineering approach is an attempt to apply such an
approach.

Computer Engineering

It might be argued that a computer is an amalgamation of **hardware** and software and hence that an engineering approach to computers as a whole might be labelled 'Computer Engineering'. In practice, the term is normally taken to refer to hardware engineering. This is usually studied within Electronic Engineering departments.

IT[1]

Programmes in Information Technology are generally quite different from those in Computer Science. Computer Science is concerned with the design of computer systems (hardware and software) as well as with their application, while IT is primarily concerned with application (although IT programmes usually include some familiarisation with computer hardware, and with how pieces of computer software work).

Once upon a time, someone invented the spreadsheet. They not only identified a need for it, but they also designed the first example of it, and implemented it by writing a computer program in a programming language. That innovative step was Computer Science. Later, other people improved the design, improved the implementation and marketed competitors. That was Computer Science, until it became routine: then it was IT. But a new innovative step would be Computer Science. The activities of using a spreadsheet, finding ways of using it more effectively, and educating other people about its use are all IT.

Computer Science is concerned with applications in which the computer forms part of what is being studied ('embedded systems'), and with systems that operate within real-time and safety constraints, as well as with the application of stand-alone computers to engineering design, science and business. To some extent, IT emphasises applications to business and administration, while Computer Science emphasises the computer system itself and a wider field of applications, but the distinction is difficult to categorise briefly in a hard and fast way: after all, you can process the results of an 'A' Level science experiment using a PC equipped with software intended for processing spreadsheets and drawing graphs – and that can be called IT.

How do you design the computer hardware and software that control a video recorder? That is Computer Science, not IT. Similarly for the computer hardware and software in a medical MRI scanner. The design of new systems for medical diagnosis, and for the automatic prescription and design of drugs, is Computer Science. But when well-established software of this kind is used by people who see themselves as medics rather than engineers, they regard it as IT, not Computer Science.

A significant development is that a wide range of powerful applications have been developed. It is now possible to do things with applications that at one time could only have been achieved by someone (a Computer Scientist) writing a program.

The distinction between IT and Computer Science is rather fuzzy and difficult to pin down. It can be argued that Computer Science includes most of IT, but it is probably true that IT excludes a lot of Computer Science. It is broadly true that IT is about applications of available systems, but that Computer Science is about how the systems themselves work.

Computer Science at university is taught as an academic subject. That is to say that, although it is oriented towards the practicalities of computer technology, it is often treated in a more abstract, higher-level manner. This involves studying some topics that you may never again encounter directly. Sometimes students think that these are a waste of time and they would rather be studying the more directly applicable topics that they see listed under IT programmes. Graduates often have a different perspective, though. They will return and say that it was only when they were tackling some interesting problem in their work that they realised why it was that they had studied, say complex **algorithms** within their degree. A degree course should train you to be adaptable and flexible. For instance, a Computer Science undergraduate may have never been taught to program in a particular language, but can soon become skilful in that language on the basis of the *principles* they were taught in other modules.

As discussed earlier, there is disagreement as to whether Computer Science is a science at all, and this is reflected in the fact that many universities house departments of Computer

Studies. On the one hand, it might be thought that the name difference is immaterial in that the modules you will study under different programmes will be much the same, but to some people the title does make a difference. While it may be more accurate not to treat the subject as a science, some people have a suspicion of any title containing the word 'studies'.[2] You will have to study the content of the modules that interest you and then decide whether the title makes an important difference to you.

A further problem is that no one agrees what all these different terms mean: what one university considers to be Computer Science is something different to another university. There really is no alternative to getting hold of the prospectus and looking at the details of the programmes. Even that is not straightforward, though. Some of the names of the modules that make up the programme may seem obvious and meaningful to you (for example, 'Programming in Java') but others may not mean a lot to you ('Algorithms', 'Numerical Analysis' or others). It may almost seem that you have to take the module in order to find out what its title means, though if you have read Part I of this book, you should understand these terms better.

You also need to think about your interests and objectives. If you are doing Computer Studies 'A' Level or Computing Highers, then you will have done some programming and you may have decided that you want to do more. You might, therefore, look for programmes with a considerable software (programming) element and avoid others which include a number of modules on hardware, for instance. Yet that might be a mistake: it may turn out that you enjoy hardware building very much once you have the chance to try it. So, unless you have a very clear idea of the kinds of topics that interest you and that you are good at, it is probably best to opt for a programme with a lot of flexibility and a number of options throughout the programme. Thus, if you discover in the first year that you do not get on well with hardware, for instance, you should be able to choose non-hardware options in subsequent years (but will not be harmed by a bit of fundamental knowledge of hardware!)

As always, talk to people who know: teachers, student guides and lecturers at open days – anyone who may be able to help. They cannot make the decision for you, of course, and you should not be unduly influenced by their personal views. Just because some student you talk to says Logic Programming is the most boring module he has taken, it does not necessarily mean you will find the same.

Beware that some of the sexiest sounding topics are also the hardest ones. The ability to generate beautiful computer graphics, for instance, may seem very attractive. Yet behind it is not so much artistic appreciation as some very hard mathematics.

One important point to get clear is that Computer Science is *not* programming. To be sure, programming is an important element and every Computer Science graduate should at least be a competent programmer, but there is much more to Computer Science than just writing programs. It is easy to make the mistake of discovering the fascination of programming (possibly at an early age) and assuming that a Computer Science course at university will be a chance to indulge that fascination for three (or four) years, but this is not the case.

The term 'hacker' is well-known but it is used in two senses. There is the one that makes it into the newspapers, when people 'hack' into private computer systems for one reason or another, but there is also the sense that is used to describe an attitude to programming. Under this meaning, a hacker is someone who throws together computer code in any way possible. It's 'quick and dirty'. They don't really care *how* it works, as long as it does work.[3] Such hackers are not the kind of person that Computer Science Admissions Tutors generally want to encourage onto their programmes – because unless they change their attitude, they will not do well on the programme.

That is not to say by any means that such people do not make it onto Computer Science courses – indeed they form a good proportion – but the person who will get the most from a Computer Science programme is someone with a deeper and broader interest in computers. Reading this book should help you to clarify these issues. You should particularly look at

Section 8.1: Computer Science, Studies, IT . . .? and Section 9.3: The UCAS Personal Statement. Talking to people will help too. It might be that at the end of your deliberations you decide that perhaps you are more of a programmer than a Computer Scientist, and then perhaps university is not the right move for you – at least not at the moment. Perhaps you would be better looking for training in programming immediately after sixth form – and being three years further up the salary ladder than your contemporaries when they graduate!

8.2 EMPLOYMENT PROSPECTS

A degree is part of your route to getting a job. A Computer Science degree is obviously a significant step towards getting a job involving computers. Given that very few jobs nowadays do not involve computers, that puts the Computer Science graduate in a powerful position. Most Computer Science graduates end up in jobs such as programmer, analyst or designer, which are very closely connected to the technology, but graduates can also end up in a wide variety of other jobs.

The greatest influence on employment prospects is the economic climate. When things are tough, it is hard for anyone to get a job, but even at such times, the prospects of Computer Science graduates are generally better than average. You will have heard frequent references to a skills shortage in the IT industry and graduates are nearly always in demand to fill these shortages.

There was a time when to have a degree was considered sufficient qualification for many jobs. However, with the widening participation in further education, employers have had to become more discriminating. You will find, therefore, that some employers will require a degree of a given class (often a 2(i); see Section 12.1: Marks and Standards).

Many universities place a great emphasis on *transferable skills*, as a means of improving the employment prospects of their graduates. It can be argued that Computer Science is more of a vocational subject and hence that most of the subjects and skills you learn are transferable. Indeed, many of the identified transferable skills taught are computer-related: word processing, use of databases, using the Web and such like. These are skills that the average Computer Science student takes for granted.

8.3   CHOOSING YOUR 'A' LEVELS OR HIGHERS

The choice of 'AS' and 'A' levels or Highers that you make will influence your eligibility for university courses. Contrary to what you might expect, 'A' level or Highers Computing is not required for most university courses. University departments are much more likely to be interested in your having appropriate qualifications in subjects such as Maths.

Entry requirements are quite diverse. Some departments will insist on 'A' levels or Highers in particular subjects, while others are not so specific about subjects and interested only in the grades you get. So, ideally you should have already started to look at the departments you might like to apply to before you make your sixth-form choices. You will keep your options most open if you take Maths and a science in the sixth form. Furthermore, for **hardware**-oriented degree programmes, it would be better if the science is a physical one (for example, Physics, Electronics or Chemistry). Of course, this assumes that you are good at these subjects, that you will want to study them in the sixth form, and that you are likely to get good grades in them.

If the truth is that you are not very good at sciences, then you should choose another subject you think you can do well in and look for degree programmes that do not require a science (ones less likely to include teaching about hardware).

If you are not very good at Mathematics, then there are degree programmes that do not require Maths at 'A' level or Highers, but beware that some of them will still require a minimum grade in Maths at GCSE or Standard grades. However, if you are not a strong mathematician, then you should perhaps stop and think about whether Computer

Science is really for you. Even if you can find a programme for which you are qualified, the kind of problem-solving, logical way of thinking that is possessed by mathematicians is also an advantage when writing computer programs. If you are not sure that you can think in that way, then you should consider carefully before pursuing a Computer Science programme at any university.

There is more information on 'A' levels and Highers and admission requirements below, in the Chapter 9: Applying to University.

8.4 CHOOSING YOUR UNIVERSITY

One point to be very clear on is that all universities are different. We talk about 'going to university' but perhaps 'going to *a* university' would be more accurate. Important differences are the value of the degree you will obtain and the experience you will have when you are there. Broadly, it is possible to distinguish five types of university in the UK, ordered roughly by age:

1. *Ancient*
   As implied by the title, these are very old, well-established universities, including Oxford, Cambridge, St Andrews and Durham.

2. *City centre*
   Universities such as Manchester, Leeds and Liverpool. These have grown up in the centre of the cities to which they are attached.

3. *Campus*
   In the 1960s, when the government wanted to increase the numbers of students gaining degrees, it spent real money and established new universities. They had to be built where land was available, so tended to be self-contained on sites just outside the city to which they are attached. Examples are

York, Keele, Lancaster and Warwick (which is nearer to Coventry than Warwick).

4. *Post-1992*
   These universities are colloquially often referred to as the 'new universities', but 'new' is rather an ambiguous qualifier (the campus universities were the 'new' universities in the 1970s and 1980s). The post-1992 universities represent another government attempt to increase the number of university graduates. This time, however, new institutions were not built (as in the 1960s); rather, existing institutions – mostly polytechnics – were redesignated as universities. Examples are Coventry, Oxford Brookes and Liverpool John Moores (note how the founders had to be rather more creative in their names as most of them were sited in cities that already had at least one university).

5. *Higher education (HE) colleges*
   There are a number of colleges which have the authority to teach to degree level (usually they do not award the degrees directly, but do so through an associated university which validates their degrees).

The value of your degree in terms of your future career depends to some extent on which university awarded it. In practice, more important is the department in which you studied, and that is discussed further in the next section. For people who do not know a lot about university, though (who perhaps did not attend one themselves), the identity of the university may be seen as important. They are more likely to have heard of the ancient universities, for instance, and so will attach greater value to their degrees. Indeed, the order of the above list is significant in this respect, but it should be stressed, though, that this is not a reliable scale. In practice, a degree in Computer Science from, say, a particular one of the campus universities may be a better qualification than one from a particular ancient university. League tables are a rather more reliable guide.

If you look on obtaining a degree as part of your career plan, then you want to attend the one which is going to give you the best start on that career – that is to say the one that is academically best – but you should also consider other factors. You are going to be spending three or four years there, so it had better also be a place where you will be happy and comfortable.

The ancient universities tend to have a slightly formidable image: that they are full of upper-class, academic high flyers and that it is hard for other types to fit in. This is not necessarily the case, and if you think that one of them offers the course that you want and that you would be suited for, then you should consider making an application.

The experience you will get in a city university is one of city life. That is to say, on a daily basis you will be mixing with other people who work and live in the city. You will also have access to its social facilities, such as pubs, nightclubs, sports watching and so on. By contrast, campus life can be very separate from the rest of the world. Some undergraduates rarely move off campus from the beginning of term to the end, working and socialising in the same small area. Others find that claustrophobic and prefer to keep a foot in the 'real world'. For them a city university may be more suitable. Post-1992 universities are often sited in cities too (often the same city as a city university), and one of their common characteristics is their size: they tend to be large. On the other hand, HE colleges are often former colleges of education (teacher training colleges) and they tend to be small. You might consider applying to one of these if you prefer a more intimate environment. Note that both the post-1992 universities and the HE colleges generally offer non-degree courses as well (such as NVQ, BTEC, and access courses), so you are likely to mix with a greater diversity of students in these institutions.

Accommodation is important too. If you go to a campus university, then the chances are that you will spend at least some of your time living in student halls of residence. So, this is another statistic you should look into: how many years are you likely to have university accommodation? In a city university you may still get a year or more in halls, but you will probably also spend some time in private accommodation. Look into how hard that is to find and how far you may have to travel.

Access to the Internet is important for students, particularly in Computer Science. Many universities offer access in student study bedrooms, so another question to ask is whether that is available. Ideally, they should also give you some idea of the cost of the connection. Of course, that assumes that you will have your own computer, and it is a good question whether that is necessary or desirable. If a department strongly encourages you to provide your own PC, then have a look at the amount and quality of equipment that they provide. Ideally, using your own PC should be an optional luxury, meaning that you can work in the comfort of your own room if you choose to.

There are further social factors that you might take into account. Is there an area of the country you would like to live in (or avoid)? How far do you want to be from home? The usual answer to that question at age sixteen is 'As far as possible', but bear in mind that once you get to university you may find that a weekend home once or twice a term is really quite enjoyable, and that may not be possible if you have escaped too far. Conversely, economics may dictate that you choose your local university, so that you can live – cheaply – at home.

You also have to decide which country you want to study in. The university system is quite similar in England, Wales and Northern Ireland, but somewhat different in Scotland. A major difference (at least at the time of writing) is that Scottish students studying in Scotland do not pay (direct) tuition fees. This is a clear, strong incentive for Scots to apply to these universities. Students from outside Scotland are liable for fees. Should they want to apply to Scottish universities, they are clearly going to face strong competition from Scottish applicants. It is not surprising that the majority of students in Scottish universities come from Scotland.

The system is also different in Scotland in that many degree programmes are available as three- and four-year alternatives.

The four-year programme leads to an honours degree, whereas the three-year degree is a 'general' one. This structure reflects a difference in secondary education in Scotland. The traditional route to university, north of the border, is to take Standard grades around age sixteen and then Highers one year later. Students following that route thus start university a year earlier than their English, Welsh and Northern Irish counterparts, but after four years of study will emerge with an honours degree. In other words, they end up graduating at the same age and with the same degree (honours) as their non-Scottish counterparts. (See below for a further discussion of honours degrees.)

Most league tables, as discussed in Section 8.5, compare departments, but the *Sunday Times* newspaper's table compares at the university level (http://www.timesonline.co.uk/). This will give you a more objective assessment.

The most important source of information at this stage is the university's prospectus. These are all available on-line, but once you have narrowed your choices down a bit you will probably want to get a paper copy. The Admissions Department will be only too happy to send you one. Again, the simplest way to order one is usually via the Web.

8.5 CHOOSING YOUR DEPARTMENT

The proposition is sometimes made that all degrees are equivalent. This is not true. To have any degree in Computer Science means that you have reached some minimum level of academic achievement, but how far beyond the minimum threshold you went will be apparent from the department in which it was taught.

League tables will give you an indication of the relative academic standing of different departments. These tables are compiled by newspapers (for example, *The Times* at http://www.timesonline.co.uk/ and *The Guardian* at http://education.guardian.co.uk/) on the basis of data collected independently by bodies such as the Quality Assurance Agency for Higher Education (QAA) and the Research Assessment Exercise (RAE). These bodies assess the quality of teaching and research in universities and departments, but they certainly do not collate their results as league tables – so the newspapers do it for them. They combine those results with other statistics, such as staff-student ratios, library and IT spending, and end up with a number. The data used and the way the data are combined varies between different papers, which is why they come up with slightly different rankings.

League tables are a reasonable starting point in choosing a department, but do not follow them slavishly. You should look at how they have been calculated and make sure that they reflect the aspects of study that are important to you. You will probably want to aim as high in the league tables as you can so that you end up with as 'good' a degree as possible. You should be aware, though, that there is an economy involved. Just as football clubs at the top of the league will demand larger transfer fees, departments at the top of their league will demand higher grades for admission. (See Chapter 9: Applying to University, for further information on admission offers.)

You should also beware of the limitations of league tables, and the aspects of a department and a programme that cannot be quantified. What is the atmosphere like in the department? What are the relationships like between staff and students? How committed is the department to teaching and research? What level of support will you get? In other words, how happy are *you* likely to be in that department; the position in a league table and the data that have been used to compile it will not answer those questions directly.

So, once you have decided on the kinds of universities that you are aiming at, then there are still academic matters to consider. You want to get on a programme that is going to suit you best. As discussed above, you want one with the right mix of modules to suit your interests, but bearing in mind that you may not really know what your interests are, flexibility and choice within the programme may be important. All Computer Science programmes are not the same. Some may have more of a theoretical emphasis, others may concentrate almost entirely on software and not cover much on hardware, and so on. As well as looking at the modules listed in the prospectus, you might also examine the research interests of the lecturers in the department. Their interests and biases will be reflected to some extent in the style of module that is taught.

One more thing you should do before you include a university in your list of applications is to read the small print in their prospectus. As you will see in the next section, you will be concentrating on finding a programme with admission requirements compatible with your projected results, but in concentrating on that, it is easy to lose sight of other requirements that might be applied. In particular, there may be a requirement for particular grades in GCSEs. For example, it might be that, although an 'A' level in English is not required for your Computer Science programme, there is a requirement for a pass at grade C or above at GCSE. This might be imposed by the department or it might be a global requirement of the university. Make sure you cover any such conditions, otherwise you may waste effort applying to a department which will automatically reject you.

Then you may look at some of the other details behind the league tables which may be important. For a Computer Science student, the ratio of computers to students may be very important. You should also find out whether there are computers that are exclusively available to Computer Science students, or whether you will be in competition with all the other students in the university. Similarly, staff-student ratios (sometimes referred to as SSRs) are important. Clearly, you will get more personal tuition and support in a university where the ratio is high.

By now let us hope you have a short shortlist. Now it may be time to visit the places on that list. This is the best way to get an idea of where you are most likely to be happy. The best time to visit is when there is an Open Day, but be aware that there are different types of Open Day. Most universities run student Open Days intended specifically for people in your position: that is, those thinking about applying for a place. Some departments also run UCAS Open Days, which are usually a chance for students who have already applied to that department (via their UCAS forms) to have a look around. Finally, there are public Open Days, which are a chance for any members of the public (such as local residents) to find out more about the university.

Prior to making your UCAS application, then, it is the student Open Day which will be most valuable to you. There will be people on hand to offer you all the information you could possibly want. Of course, they will also be trying to sell themselves, so you may reach a slightly distorted view. Lectures are often cancelled on Open Days and students may spend the whole day lying in the sun, drinking beer and listening to loud music. Do not be fooled that this is a typical university day.

Degrees come in different flavours. Given the earlier debate about whether Computer Science is really a science, most universities offer Bachelor of Science (BSc) degrees in the topic. However, you may find some that offer Bachelor of Arts (BA) degrees. The difference is really irrelevant. There is also a Bachelor of Engineering (BEng) degree, offered by departments which consider Computer Science to be more of an engineering discipline, but again the designation is not very important.

Bachelors degrees are generally of three years' duration in England. This will usually lead to the award of a degree *with honours*. There used to be a clearer distinction between an *honours* degree and an *ordinary* degree, but that has largely been lost now, at least outside Scotland. Most degree programmes are aimed at the award of honours degrees. If a student does not get a sufficiently high mark at the end of their studies, though, they may be awarded an ordinary degree. In other words, 'ordinary' means they nearly failed, but not quite (see also Section 12.1: Marks and standards.)

The system is rather different in Scotland. There, as mentioned earlier, a three-year programme leads to the award of a general degree, but most programmes are four years long and lead to an honours degree. This reflects the difference in the secondary education system in Scotland, where students normally spend just one year in the sixth form, take Highers exams, and then go to university one year earlier than their English and Welsh counterparts. In other words, most students in the UK end up graduating at the same age with the same kind of degree. Scottish programmes are also different in that the first year is generally treated as more of a foundation: students study a number of different subjects during that year, only a portion of which will be Computer Science.

Some departments outside Scotland also offer four-year programmes, but these lead to Master of Engineering (MEng) degrees. An MEng degree is a masters level qualification (although not all employers recognise it as such – they expect a masters-level graduate to have a bachelors degree as well). What is more important is that an MEng degree carries a higher level of professional recognition. A professional engineer is recognised as such by becoming chartered, and the first step to achieving chartered status is to obtain an accredited MEng degree. With a bachelors degree, you have to undergo further training before you can start on obtaining chartered engineer status. So, if you see your future as being a professional engineer, you should seriously consider registering for an MEng course.

Courses (bachelors and MEng) may be accredited by professional bodies, the British Computer Society (BCS) and, in a small number of cases, the Institution of Electrical Engineers (IEE). That means that the course has been through yet another round of quality control, ensuring that it meets the appropriate engineering standards. Again, if you see your future as a professional engineer, then you ought to look for such accredited courses.

Some universities offer sandwich variants of their programmes. There are different kinds, but they will usually involve students spending a year (usually after the second year of the programme) working in industry. This can be a most valuable experience. It gives you a taste of what it is like to work in industry and it can help you to refine your ultimate career objectives. Not least, it can give you experience of the kind of work that you might like to do in the long term – or that you might prefer not to do! Also, when you do graduate and are looking for a job, your placement may give you the edge over a straight Computer Science graduate who does not have that practical experience. The only disadvantage of taking a sandwich course is that you will not graduate for another year.

An alternative way of getting industrial experience is to find a summer internship – that is, you spend one (or more) of your summer vacations working in a computing-related job. The availability of internships – and sandwich placements – depends to a great extent on the state of the economy at the time. In hard times, companies will suspend sandwich placements, but even in such times they may be open to the shorter-term commitment of internships.

Some departments will be better organised than others in giving assistance in finding and administering sandwich placements and internships. If they are an important component of your Computer Science programme, then you should find out how much support each department is likely to offer.

NOTES

1. Most of this explanation is derived from that eloquently expressed by Bill Freeman on http://www.cs.york.ac.uk/admit/Differences.html#csit
2. There is an instance of an American computer manufacturer which was very keen to have its products in use in a British university Computer Science department. They were already installed in a Computer *Studies* department, but the company was not happy with that and offered a very good deal to get them into a Computer *Science* department.
3. The two definitions of 'hacker' are not unrelated. It is the same absorption with solving problems and getting round obstacles that drives both kinds of hackers – and, of course, they are often one and the same person.

9 APPLYING TO UNIVERSITY

The choice of university is not a one-way process and you do not simply decide where you will go – the university has to choose whether they are going to accept you. The mechanics of applying and the selection process are explained well on the UCAS website, www.ucas.com, but in this section we can give you some ideas that might help you to get onto the Computer Science programme of your choice.

All applications are made through the UCAS form. Much of the form is quite straightforward to complete: it is simple factual information regarding you, the schools you have attended and the grades you have attained in the exams you have already completed (for example, GCSEs, 'AS' level, Scottish Standard grades or whatever). You can specify up to six universities on your UCAS form. It is to be hoped that, following the advice in this book, you will end up listing six universities, any one of which you would be happy to attend, with a good chance that more than one of them will want you. You should be aware that the admissions tutor in any one of them will not know which other ones you have applied to. Before UCAS hid this information, there were rumours that if you applied to University X, then University Y would not even look at your application.

This part of the book is largely about finding the 'right' six universities to list on your form, but it also looks at two sections of the form which can influence your chances of getting in to the university of your choice, and improve the probability of *them* wanting *you*. They are the personal statement and the reference from your school or college. Section 9.3 gives some advice on composing your personal statement, but first we will look at how to decide which six universities and programmes to put down as your choices.

Departments usually have a 'standard offer': that is, the exam grades – or Tariff value (see below) – that they normally require of students for admission to their programme. This standard offer is usually openly publicised in the prospectus and elsewhere. It is important to be aware that standard offers are part of a simple market economy. Furthermore, Computer Science remains a popular topic, with more applicants than places. A department with a lot of places to fill will make lower offers than one with fewer places. To be more precise, a department that generally gets a high proportion of applicants to places will have a higher standard offer.

This is important because applicants often confuse 'price' with quality. In other words, there is an assumption that a programme with high entry requirements is better than one with a lower standard offer. This is not necessarily true. There is at least one instance of a Computer Science department which was oversubscribed with applicants. It raised its standard offer, hoping to deter some people from applying, but in practice it had the opposite effect. Evidently students thought that 'you get what you pay for' and that the programme had somehow improved because it had raised its admission requirements!

The first thing most admissions tutors will look at on your UCAS form is the predictions your teachers have made as to the grades you are expected to achieve in your end-of-school exams. Clearly, to have the broadest choice of universities, you want the highest possible predicted grades. There is not much you can do to influence these predictions, apart from working hard and doing as well as you can in your sixth-form work.

In making your application, though, you should be aware of your predictions. If you are predicted three grade Cs in your 'A' levels and the standard offer of one of the universities which interests you is three Bs, then you have just wasted an entry on your UCAS form: the admissions tutor is unlikely to look any further at your application.

Having identified all the best academic departments to which you might apply, you might want to also look a bit further down the league tables. Highly-rated departments are likely to have high requirements in terms of the grades they will require you to obtain. While you will, of course, hope to achieve such high standards, things can go wrong. In order to try to ensure that you end up with a place at a university, you might think about including one or two universities with lower standard offers, as a form of insurance.

If you are predicted the right level of results, then the tutor will decide whether to make you an offer or not. It is not possible to say exactly on what basis they will do this. Very few departments interview candidates, so they are relying entirely on that which is on your UCAS form, in your personal statement and in your reference. What they are looking for is someone they think will do well on the kind of programme they offer, someone they expect to graduate in three or four years' time with a good class degree. Doing your research into courses and selecting those that you think will suit you should help you on the way.

9.1 THE ADMISSIONS PROCESS

Universities and departments have a tricky numbers game to play. They have a quota of admissions places that they are expected to fill, and if they miss that quota – going over or under – then they suffer financial penalties. It is under these conditions that admissions tutors make offers to prospective students without knowing what those students' grades will be. Yet all decisions must be made on valid academic grounds.

In principle, a candidate cannot be turned down merely because the tutor believes that sufficient offers have already been made to achieve the quota. Nevertheless, if you want to maximise your chances of getting into the department of your choice, you should submit your application as early as possible. Copies of your application will be sent to each of the universities to which you have applied. They will decide whether to make you an offer of a place.

Unless you have already completed your 'A' levels or Highers (perhaps you took a gap year or two), then the offer will usually be conditional. That is to say that they will offer you a place if you achieve the prescribed grades (remember, most departments publicise their 'standard' offer). This may be expressed quite specifically (for example, for someone taking 'A' levels, they might specify an A and two Bs, and the A must be in Mathematics) or more generally, in terms of the UCAS *Tariff*. The Tariff gives you a number of points depending on your 'A' levels or Highers, but takes account of a wider range of other qualifications as well. Some of the common qualifications and their Tariff points values are given in Table 9.1. Other exams such as BTECs and music grades may also be included. More details of the Tariff can be found on the UCAS website, at http://www.ucas.com/candq/tariff/

Points	England, Wales & Northern Ireland			Scotland	
	'AS' level	'A' level	Standard grade	Higher	Advanced Higher
120		A			A
100		B			B
80		C			C
72				A	D
60	A	D		B	
50	B				
48				C	
42				D	
40	C	E			
38			Band 1		
30	D				
28			Band 2		
20	E				

Table 9.1 UCAS Tariff points values for different UK grades. This is only a sample. Other qualifications are included in the table at http://www.ucas.com/candq/tariff/

Using the Tariff, entry requirements might be expressed as follows: 'Offers will normally be in the range of 160–220 points including a minimum of 140 points from two or three full six-unit awards (A2/Advanced GCEs or VCEs) or one full twelve-unit award (double VCEs or BTEC NC/ND) or one full six-unit award plus two three-unit awards (AS Levels).'[1]

Scottish students will take Highers and have the opportunity to take Advanced Highers. Of course, the major difference between 'A' levels and Highers is that Highers are taken after one year of sixth-form study. Most would-be Computer Science students will take Advanced Highers. They will thus apply to universities after they have completed their Highers and will know their results already.

The admissions requirement for Computer Science in most Scottish universities is in terms of Highers grades; they tend not to require Advanced Highers. On the other hand, English universities often do ask for Advanced Highers (presumably because they are more similar to two-year 'A' levels with which they are more familiar – and remember that English programmes are generally a year shorter than Scottish ones). Notice that the entry requirements are almost invariably expressed in terms of grades and not Tariff values, possibly because the Tariff values assigned to Scottish qualifications are seen as anomalously high. Thus, it is likely that any offer from a Scottish university will be unconditional – based on the already-known Highers results – but any English offers will be conditional on the Advanced Highers results.

Qualifications are also accepted from other countries. Most departments will have a standard offer for the International Baccalaureate (IB). Overseas applicants are also likely to require a suitable English language qualification, such as IELTS (International English Language Testing System) or TOEFL (Test of English as a Foreign Language).

You can hold at most two offers at any time. The ideal is to hold one from a department to which you really want to go and a second, lower, 'insurance' offer from another department, just in case you do not get the grades you hope for (and were predicted). If you are Scottish, and you have an unconditional offer from a Scottish university, then you might hold a conditional offer from an English one as your insurance.

In August, the 'A' level and Highers results are released. Unfair as it may sound, the departments for which you hold offers will know your results before you do. Thus, on the day you get your results, they will already know whether you have achieved their conditions and whether you will be getting their place.

In the unhappy event that you have not achieved the grades, then all is not lost. Firstly, if you are just a point or two below the offer from the department of your choice, then they *may* still accept you. Remember, they have a quota to fill. If there are a number of students like you who have not achieved their predicted results, then they will be in danger of under-shooting that quota and they are likely to add in the best of the near-misses.

In this situation, you should try to contact the department by phone as soon as you can. The phone lines will be busy, but keep trying. It is worth speaking to the admissions tutor if you can. Remember, though, that this is a stressful time for admissions staff too. When you finally get through, even if you have been pressing redial repeatedly for the past half-hour, be polite and cheerful: try to sound like the kind of person they would want in their department! Explain the situation, that you have narrowly missed their offer and ask whether they may be able to accept you.

If that does not succeed, you can resort to the Clearing process. This is where departments that have not achieved their quotas (and in the current climate, there will be many of them) try to grab students who have not achieved their grades. You will now be under a time pressure that you were not under when you were making the selections for your UCAS form, but you should try to be as thorough in assessing departments: you are still about to commit yourself to spending three or four years there and to emerge (you hope) with a degree with that university's name on it. Look at the departments offering places, and then go back through the process outlined in the previous sections on choosing a university and a degree. In making these choices, beware of admissions tutors who seem too keen to have you: why have they fallen so short of their quota?

Not achieving the grades you needed can be a problem, but what if you do *better* than expected? Suppose you are predicted mediocre grades, you therefore do not get offers from the university of your choice, and you end up accepting an offer from a university lower on your list of preferences (and keeping an even less desirable offer as your insurance choice). But then you surprise your teachers by doing better than they expected – indeed, you might even get grades better than the standard offer from your first-choice university which has turned you down. If you think you really can do better than the offers you are holding, then you first have to get released from them. If those departments agree to that, then you show up in the UCAS records as 'rejected', and you can then enter the clearing system in the hope of finding a better offer. Alternatively, you may withdraw from the system all together and then re-start in the UCAS system for entry in the following year.

9.2 GAP YEAR

What if you want to take a gap year between school and university? The best course of action is to apply for a *deferred entry*. That is to say, if you are taking your 'A' levels in the spring of 2007, you would submit your UCAS form as usual in 2006–07, but you would state that you want to defer your entry until 2008. Your application will be processed along with all of those for entry in 2007, but in the knowledge that you want to be part of the 2008 intake.

You might think that it would be better not to apply until after you have completed your 'A' levels, that you could then have a year off and a bit longer to think about your choices of universities. However, if you think about it, you do not really have a year out, that way. You would have to be filling in your UCAS form in autumn 2007 and going through the admissions process over the next few months. That might hamper whatever it is you plan to do in your gap year. It might be very inconvenient if you have to make a decision about an offer from a university, when you are backpacking in Australia and hard to contact!

Generally, it is better to use the deferred entry scheme. You will receive any offers in 2007 and when you find out your 'A' level results you will know if you have met any conditional offers. Then you can go off on your world tour or whatever you want to do, knowing that you have a secured place to come back to.

9.3 THE UCAS PERSONAL STATEMENT

As suggested above, there are two main requirements to get an offer from the university of your choice. First, you have to achieve good enough exam results to meet the department's standard offer. Then, you have to make yourself stand out in some way from all the other people who have applied to that department who are also predicted to achieve the required grades. This is where the personal statement on the UCAS form comes in.

You may think that the way to catch the eye of the admissions tutor is to start your personal statement with a sentence such as 'I have been fascinated by computers since the age of seven when my father first brought a PC home from work.' You won't. First, it is not original. Nearly every form submitted for Computer Science programmes starts with something like that. Secondly, it gives precisely the *wrong* message. As discussed earlier, in Section 8.1: Computer Science, Studies, IT . . .?, this will probably mark you out as a hacker, a tinkerer with the technology, which is not necessarily the kind of student Computer Science departments want to encourage.

What admissions tutors (in all subjects) look for in a personal statement is evidence that the applicant is:

- well organised
- independent
- motivated.

Most admissions tutors and lecturers probably think that the best kinds of students to have on their courses are people like themselves, so that is what they will look for in a personal statement. Of course, you probably do not know the people reading your UCAS form, but you might like to try to create an image of them while you are composing your personal statement. He (and the bald truth is that the chances are very heavily weighed that he will be male, see Section 9.5: Are Girls Allowed?) is not a hacker. He has an interest in broader and deeper topics around Computer Science: Maths, Hardware Design, Human-Computer Interaction, and so on. Try to show him you have an interest in one or more of these areas, that you have thought about them and have an opinion about them. You do not have to have the same opinion as the UCAS form reader – just one that you would be willing and able to defend.

The person reading your personal statement will also have interests outside his job. There are no published statistics, but anecdotal evidence suggests that a very high proportion of Computer Science lecturers play musical instruments. So, if you have an interest in music, say so. If you do not play music, you should still mention your Duke of Edinburgh awards, or your voluntary work with old people, or your antique collection – or whatever else you do that makes you a well-rounded person, the sort of person who is going to make interesting contributions in tutorials.

Of course, the personal statement is your chance to sell yourself and you should use the most positive language you can, but try not to be too clichéd. Describing all your extra-curricular activities as 'challenging' is almost meaningless. Think about what *you* really got out of your Saturday job, or completing your Duke of Edinburgh. Listing your extra-curricular activities not only shows how well rounded you are, but that you already have experience of time management, an invaluable skill for a university student.

As a Computer Science undergraduate, you will be expected to work hard. The admissions tutor may be reassured if it is apparent from your personal statement that you are aware of that and that you are already used to getting down to hard work. You might think it sounds impressive that you are going to get good exam grades with little effort because you are naturally gifted, but it is going to be more appealing if you show that you are accustomed to hard work. It is probably more believable too. The impression you should be able to give is that you work hard on your academic studies, but that you also manage to make time to take a full part in your extra-curricular activities.

If you have followed the process advocated in this book in selecting the programmes for which you apply, then you should have selected those which match you, your interests and abilities. Try to bring this out in your personal statement. For instance, if you have chosen programmes with a high mathematical content, then make it clear that this was a deliberate choice, that you enjoy the Mathematics you are studying. Alternatively, perhaps you have chosen programmes with a **hardware** element, even though you have not had much experience of hardware construction. In that case, make it clear that this is a new field that was not available to you in school or sixth form and that you want to study.

Writing your personal statement is a good chance to stand back and ask yourself some important questions. Not the least, why do you want to go to university? If it is part of a clear plan you have for a career, then you should get that across in your personal statement. That will be much better than giving the impression that you are applying to university because you have reached the end of the sixth form and you cannot think of anything better to do. Your career plan does not have to be pinpoint sharp and set in stone, though. You might say that you think that you would like a career in, say, software development, but you are not entirely sure in what role (programmer, systems analyst, or whatever). It is perfectly acceptable and realistic that your ideas as to what interests you specifically will become clearer during your university career.

Having composed your personal statement, taking all the above into account, please have it checked carefully. Getting the grammar right – apostrophes in the right places, and so on – will make a positive impression. Do not underestimate the importance of this. It is tempting to think that you are not applying to study English and so the people who read your form will not be concerned with (or even very knowledgeable about) the nuances of English grammar. However, you should remember that computer scientists write programs, and programming languages are *very* strict about their grammar – where you can and cannot put a semicolon, for instance. Therefore, Computer Scientists and those who deal with admissions to their programmes are likely to be very fussy about quality of writing. If you feel you may need help to get these things right, and you have not spoken to your English teacher since before you took your GCSE, then Truss (2003) presents many of the requirements of correct English in a readable form.

9.4   REFERENCE

The other free-text section on the UCAS form is the reference from your school or college. That will be written about you by your tutor or tutors, so there is not much you can do about its content – other than impressing your tutors! Yet that is not a silly suggestion. When your UCAS form appears on your tutor's desk, is he or she likely to know immediately who you are and what to say about you, or are they going to have to go and pull out your student record to find out about you? The point is that if you are clearly a highly-engaged student, they are going to find it easier to write a positive reference for you. And no one but you can influence the impression you make.

9.5   ARE GIRLS ALLOWED?

It is an undeniable fact that Computer Science departments are dominated by men, both in the academic staff and the student bodies. It is beyond the scope of this book to speculate as to why this is so, or what might be done to redress the balance (if you are interested, you might look at the *Women into Computing* website, http://www.wic.org.uk/). If you are female and contemplating a Computer Science degree, then you should be aware of various facts.

Positive discrimination is illegal, so you cannot be given any more favourable treatment on admission. However, if you are a strong candidate and you happen to be female, you can expect a good offer.

You may want to ask about the gender ratios in the departments that interest you. For the students, the statistics will probably be on the department's website. If you think you would be uncomfortable in a lecture with 100 males and only one other female, then you should look around for departments with better ratios. You might also think about programmes that allow elective modules from other departments, so that you can choose modules in which the gender ratio is likely to be more balanced – or reversed. Of course, you do not spend all your time in lectures, and you may feel able to put up with the gender ratio as long as you are in a university in which you can socialise with other women and mix with people – male and female – from other courses.

The university website may not give the same gender figures for staff. From prospectuses and websites it may not be obvious how many female staff there are in a department. While some departments will list first names, from which you can deduce who is female, others present names in a gender-neutral way, using initials, rather than names. Where titles are used, most lecturers will be 'Dr', so that does not help either. Do not be fooled by the accompanying photographs of scenes in the department, either: departments keen to attract more female students will take care to ensure that women are well represented in any pictures they publish.

Remember that you will be assigned a tutor from the academic staff, who is someone you may have to discuss personal problems. In a department with a higher number of female staff, there is a better chance that your tutor will be female. If you think that may be important to you, you should ask for a female tutor, though it may not always be possible to meet such requests.

9.6   I HAVE A DISABILITY

Under the Disability Discrimination Act (DDA), no university is allowed to discriminate unreasonably against any student, or prospective student, on the grounds of disability. Having a disability should not, therefore, be any bar to your studying Computer Science. Of course, 'disability' covers a wide range of conditions. It is beyond the scope of this book to go into them all with regard to their effect on anyone studying Computer Science, but a few points should be made.

Whether you have a disability or not can be hard to ascertain. If you have had a Statement of Special Needs (a *Statement*) while you were at school, then clearly you do have some form of disability that affects your education, and you will need continuing support in higher education. If you were not statemented in school, you might still be classed as having a disability in higher education. The onus is on you to take the initiative to find this out. If you have – or think you may have – some form of disability, then it is worth pursuing it in order to ensure that you receive whatever support is appropriate in your university career.

In practice, Computer Science students with disabilities probably have better opportunities than those studying other subjects because computers have become so flexible and adaptable for access by people with all forms of disability. It is not unusual for a person with a disability to start using a computer as a form of assistive technology – as a means of communication, for instance – and for then he or she to become interested in the technology itself and to want to study it in more detail, perhaps to the extent of taking a degree.

Statistically, the disability you are most likely to have is dyslexia. The effects of that condition vary greatly between different people. It is easy to think that Computer Science is more concerned with the writing of programs than of essays, and that the amount of reading expected will be much less than for, say, a History student, but it is not always as simple as that. Some of the effects of dyslexia can have a profound effect on people's ability to study Computer Science. At the same time, anyone who has got as far as applying to university despite their being dyslexic has devised good strategies for coping with the effects of the condition and should be able to continue his or her studies with the support available.

Most universities are well aware of the incidence of dyslexia and have mechanisms in place to assist students, although the support will be better in some universities than in others. Beware of the attitude that 'You can have 10% extra time in exams – and that's it; your special needs have been met'.

If you have a condition that affects your mobility, then you will have to think about physical accessibility. Again, the DDA mandates that access to buildings should not be a reason to exclude anyone from studying in any particular university. So it is up to the university to ensure that you have the access you need to its facilities. Clearly, this will be easier in some universities and departments than others. Those in new, modern buildings probably have physical access built in. Those in older buildings may not be able to ensure your access to all rooms, and will have to accommodate your needs in other ways. You may find that you have to have tutorials moved to ground-floor rooms and have all your lectures specifically timetabled to be in accessible rooms. You have the right to study in the department of your choice as long as you meet its admission criteria, and if the department happens to be in an old, inaccessible location, that is somebody else's problem.

There is almost no form of disability that cannot be accommodated with regard to computer access. Blind people can use screen readers that render the contents of the screen in speech and/or braille. People who cannot use a keyboard and/or a mouse can use alternatives such as speech input, and so on. In fact, for many Computer Science students with disabilities, it can be the non-computer-based parts of the course that are most difficult: access to lecture material, getting books in an accessible format, and so on.

Disabled students in higher education are entitled to financial support to cover the cost of any extra help that they need.

A Disabled Students' Allowance (DSA) can cover the cost of special equipment (such as a laptop computer for note-taking or adaptations to existing equipment), personal help from people (note-takers, personal assistants and the like) and other extra costs. To obtain a DSA, you must apply to your local education authority (LEA). They will require you to have an assessment of your study needs. This can be carried out by a body such as the *National Federation of Access Centres* (http://www.nfac.org.uk/) or *AbilityNet* (http://www. abilitynet.org.uk/). More information on obtaining a DSA can be obtained from *Skill*, the National Bureau for Students with Disabilities (http://www.skill.org.uk/).

Any university you apply to will have a disabilities officer, and most departments will have someone with specific responsibility in that area. It is important that you liaise with them as early as possible. You and they need to know precisely what your needs will be. Beware that whereas you may have a computer system at home that you can use very well, it may not be so easy on a course. Which operating systems will you be expected to use on the course? While you may have developed very accessible ways of using, say, Windows, you may be required to use Linux during your studies.

Although the DDA protects you from unfair discrimination, be aware that it does not mandate miracles. A university must take *reasonable* steps to accommodate your needs. What is 'reasonable' is, of course, open to interpretation. A department might argue that it is not reasonable to expect the whole of a new textbook to be available to you in braille for the first week of a new course. If you think at any time that you are not being treated reasonably, then there are several steps you might take. First, your tutor should be able to help. The next step might be to talk to the Students' Union, who will have a disability officer or welfare officer. If you cannot resolve the matter within the university, then you might consult Skill. Ultimately, any dispute over disability access might be taken to the Disability Rights Commission (http://www.drc-gb.org/).

The UCAS form has a section in which you can declare any disabilities, but it might be that you prefer not to do so.

However, if you do not declare a disability, then it may be that you lose some of the protection of the DDA. If a department has no reason to know that you have a disability, then it cannot be censured for not accommodating your special needs.

Reading Part I of this book should have given you an idea of what Computer Science is. If that motivated you to go ahead with an application, then Part II should have helped you through the application process. Once you have completed that process, you will hope to have a couple of good offers in hand. Then you just have to get the required grades in your exams. That's all.

Assuming you achieve the grades, you're on your way to university, and the final part of the book will help to prepare you, to settle you in and to help you do well in your university career.

NOTES

1. Taken from the University of Wolverhampton's website entry for Computer Science (http://www. wolverhampton. ac.uk/).

PART III
Study Skills

10 EAT WELL, WORK HARD AND GET TO BED EARLY

This part is meant to give you an idea of how you can expect to work on a Computer Science degree course. It is not intended to be a comprehensive study skills guide; rather, it sets out some of the things that are different in the life of a Computer Science undergraduate compared to other students.

The fundamental study skills for a Computer Science student are much the same as for any other undergraduate. You do have to work, but you should make the most of your time. Many people believe that the maxim 'Work hard, play hard' has particular applicability at university. You are unlikely to have such a degree of freedom again. So, the freedom is something to enjoy. The problem is that some people cannot cope with that amount of freedom, they forget about the contingent responsibility to study and end up curtailing their university career by failing exams and being asked to leave.

Time management is thus a vital skill to develop. In some ways it can be the most important skill that you will learn. An undergraduate degree programme is probably the most unstructured environment that you will work in. There are a large number of graduates with 2(ii) degrees who had the academic, intellectual ability to get a 2(i), but who were not able to settle themselves into the necessary work pattern.

When you start on your degree programme you will be looking forward to being a graduate, but you may have little thought about what class of degree you hope to end up with (see Section 12.1: Marks and Standards). However, it is probably worthwhile giving some thought to it as soon as possible. In particular, if you believe that getting a high (first-class) degree will be important to you, then you will have to pursue it from the start. A few graduates with first-class degrees achieve them with little effort, through pure natural ability, but the majority of firsts are obtained by very hard work. So, if you want to aim for a first, you must keep on top of the work all the time. To do so, follow the advice in this book and in other study guides (such as Becker and Price 2003, and Northedge, Thomas et al. 1997). An important point to remember is that you have to sustain the effort. There is a well-known phenomenon of the 'second-year dip'. That is to say, students of all levels often get worse results in the second year that they did in the first. There are many possible explanations of this phenomenon, but they all probably amount to people easing off because they have survived their first year and feel a long way from graduation. Clearly, it is worrying for anyone to have a decline in their marks, but if you are looking for as good a degree class as you can get, then it is definitely to be avoided; you are going to have to work at least as hard in your second year as you did in your first. (See also Section 12.1: Marks and Standards below on how your marks will contribute to your overall result.)

A likely consequence of a decision to work for a first is that your social life will be curtailed: you need to spend a lot of time on your studies. That is a choice you have to make when you consider what you want out of university.

Undergraduates have lectures and labs at prescribed times. However, attendance may not be compulsory, and this is the first management decision you will have to make. There may be times when you judge that your time would be better spent working on an assessment than attending a lecture, and there will be times when you are tempted to judge that another hour in bed is more valuable than a lecture. It will be up to you to decide.

You will be set optional work: reading or exercises that will not be marked. Again, it will be up to you to judge whether it is worth spending your time on them – bearing in mind that if the lecturer has thought it worthwhile setting the work, then it probably is worth doing it. For some pieces of work, though, there will be very definite, hard deadlines. Coursework will have to be handed in on time and it is up to you to make sure that you meet that deadline, along with everything else you have to do.

A social life and academic work will probably not be all that you will have to fit in. Many students decide that it is necessary to obtain part-time paid work to support their finances. Again, you will have to apply your judgement; how much time can you spend on jobs without affecting your studies? Similarly, you will have to work out when you can work and when you can study.

Time management is undoubtedly a vital skill and it is certainly a transferable skill. If you can control and apportion the time you spend on studying, then you can have the 'spare' time to pursue the other opportunities that university offers – whether they be sports or culture or whatever else interests you. Again, university is a unique chance to try those things with special student rates and societies to support you. So, if you think you might like sky-diving, try it.

The timetable of a Computer Science student tends to be quite dense. If you share halls with History students, you might find it irksome that they may have just three hours of timetabled work per week, while you are working almost full-time days, Monday to Friday. You have to ignore those differences, though. What you do not see is the amount of preparation and reading they have to do in order to prepare for those three hours of seminars.

One of the hardest bits of time management is filling the little gaps. There will be times when you have an hour between one lecture and another. It is very easy to fill that time in a coffee bar, but that could be one hour of very productive work.

11 TEACHING AND LEARNING METHODS

11.1 LECTURES

Lectures are still the basis of most modules. Lecturers differ greatly in style and presentation. Some still use the blackboard and 'chalk and talk', but many Computer Scientists are comfortable with the technology and will use computer projection – what is becoming known as the 'PowerPoint' lecture. You should not be seduced by the technology. Software (such as PowerPoint) makes it easy to produce something that looks slick, but it may be wrong or content-free. If the lecture slides are computer-generated, though, there is a better chance of your being allowed access to your own copies. This can relieve you of the chore of taking notes, though that is not necessarily a good thing. Taking your own notes as the lecturer talks may help you to understand and remember what has been said. Whatever technique the lecturer uses, try not to just sit there and passively let it wash over you.

One reason lectures still occur is because they are a chance for interaction. A lecture need not – and should not – be a one-way communication. *You can ask questions*, and most lecturers are only too happy to receive them. A question is a sign that someone is paying attention. It is also feedback. Without any 'audience reaction', lecturers can only assume that everything they have said has been understood and that – if anything – they ought to be increasing their pace.

Students are usually very inhibited about asking questions. People fear it is a way of displaying their ignorance in a room full of people. In practice, if you have not understood something, then the chances are that there are several other people in the lecture with the same doubt. If you ask the question, you are doing them all a favour. Sometimes students prefer to approach the lecturer at the front at the end of the lecture. That is less public and gets the question answered, but then all the other students miss the benefit of the answer.

After a lecture, there is a temptation to file your notes (hand-written or handed-out) to be revisited at 'revision time'. That is really not the best idea. You should read through them after the lecture, and you should make very sure that you understand them. If there is anything you do not understand, do something about it. For example, you could read the text-book, or ask friends if they understood. If that fails, then ask the lecturer. Different departments have different levels of accessibility. If you are lucky, you can drop in on the different lecturers and ask them if they have five minutes (but don't be upset if they say 'No'; they don't spend all their time between lectures waiting for students to call by!). You may have to make an appointment (email is often a good way), or you might email your question to them.

Reading the textbook is a good idea anyway. Some lecturers will explicitly tell you which parts of the text to read for each part of the course, but if they don't, then they are probably assuming that you are reading along anyway. Do not expect to be able to complain about the exam question on a topic that was not covered in depth in the lecture if it was covered in detail in the text.

11.2 LABS

Computer Science is a practical subject and many topics within it can only be taught in a practical manner. Many modules have a practical component involving exercises to be completed in a computer lab. There will generally be help available from the lecturer and/or graduate student teaching assistants. Like lectures, the lab is a chance to put your hand in the air and get some assistance – and as with lectures a lot of students are reluctant to do this. It is an opportunity you should not waste.

It should be clear what the objectives of a lab-based module are. It may be that there is sufficient time scheduled for you to meet the objectives in the timetabled sessions but, if not, it may be necessary for you to spend some extra time in the labs to complete the work.

Most Computer Science programmes will include software labs. Some will also include hardware lab sessions. They are run in much the same manner, but obviously the objectives will be rather different: the building of some piece of hardware.

11.3 EXERCISES

Some modules include regular exercise sheets. The marks for these may or may not count towards the overall mark for the course. Either way, it is in your interest to treat them seriously and to put some time and effort into attempting them. Solutions may be published (usually some time after the questions are set). Do not be tempted to look straight at the solutions. It is easy to convince yourself 'I could have done that', when you see the solution; the only way really to know is to try for yourself, blind.

11.4 NOVEL TEACHING

It may be that some of your modules do not use – or do not rely on – the traditional teaching methods of lectures and such like. Increasing use is being made of IT, in particular. Currently this is often based on the Web, whereby the module has a Website with related materials and other facilities such as chat rooms and notice boards for communication with the other students and lecturers. You may use other computer-based tools too, such as simulators or software that will automatically mark programming assignments for you.

One thing that you can expect – despite increasing numbers of students and static levels of funding – is that new forms of teaching technology will continue to be introduced where possible.

11.5 REVISION

The idea of revising for exams is possibly somewhat outdated. Traditionally, university courses involved three years of study which was assessed in final exams at the end of the course. Clearly, to be assessed in year three on work one had done in year one, it was necessary to go back and remind oneself of the material. In modern university programmes, exams generally follow the modules to which they relate(although some of the older universities still use the more traditional approach).

Being assessed immediately following the teaching means that there is generally less time for revision, but it also means there should be less need for it. If you follow the material as it is taught, go to any labs, read around the subject, and do any exercises, then there will not be any need for revision; you will understand the material and it will be fresh in your mind. However, timetables may leave provision for revision weeks and you can still make good use of that time. Rather than revising as such (that is, reminding yourself of the material) you might concentrate on exam preparation – practising on past papers, for instance.

11.6 PROJECTS

Project work is an important part of most programmes. There are generally two kinds of projects: those associated with particular modules, and stand-alone projects, assessed in their own right.

Some modules may be based largely on a project. That is to say, there may be lectures and other supporting teaching, but the main objective (and assessment) is based on the completion of a project. The project may be undertaken individually, in small groups (perhaps pairs) or in larger groups (even the whole class). Obviously this requires different skills and methods to those needed in a conventional module with an exam at the end of it, and it is more similar to the kind of work one would do in a job.

Most Computer Science programmes include a stand-alone project. This usually occurs towards the end of the programme, and hence is often called the 'final-year project'. The idea is that – to some extent – the project is the culmination of the study programme: take all the things you have learned throughout the programme and apply them to a realistic problem. Projects are often undertaken by individuals, though there may be some level of collaboration and teamwork. The project will normally be supervised by one of the lecturers and you can expect to meet the supervisor regularly over the duration of the project. A project will normally have an extended duration, and it may take anything up to a whole academic year.

You can expect help in choosing a project topic. Lecturers will usually publish lists of ideas for projects from which you can select. Some departments will also allow you to define a project of your own. In that case, you will have to find a lecturer willing and able to supervise your project. There will generally be more work involved in creating a self-defined project, and you have to be careful to define one that has sufficient academic challenge but is not over-ambitious – given the time and the resources that you will have available. On the other hand, if you do define your own project, then it should be something that really interests you and that you are therefore likely to do well. Regardless of who defines the project – you or your supervisor – it is important that the topic is something that truly interests you because you are going to spend a lot of time on it and it will contribute a significant amount to your result. Given that project work is the activity most similar to what you will be expected to do in a job, departments often place a lot of emphasis on projects.

Sometimes projects are undertaken in collaboration with outside bodies, such as small businesses. It can be very rewarding to know that someone is really interested in the outcome of your project – to the extent that they may use it to solve real problems. However, there can be downsides to such projects too. First, you (and your project supervisor) need to be sure that there is sufficient academic challenge in the project. A common requirement in small businesses is for a database of some sort (for example, for tracking stock). Clearly, databases are an important contribution that IT makes to modern business, but to implement another stock database is not particularly novel. If you want a more research-orientated project (see below), then a database implementation will not be very suitable. Secondly, you should also be careful that you are not simply used as 'cheap labour'. The company commissioning your project has its own objectives and you must make sure that they do not deflect you from the academic objectives of your project work.

Projects can take a wide variety of forms, but in Computer Science they commonly involve building something – usually a piece of software. Just building something is not enough, though, and there is generally a need to reflect on what you have done and to evaluate whatever you have built. An important part of the project is writing it up in a report. You might also be required to give a presentation and/or a demonstration of whatever you built. These might form part of the assessment.

Projects can usually be placed on a spectrum, as depicted in Figure 11.1. At one end, there are projects that might be characterised as 'development'. That is to say, there is nothing particularly novel about them. They probably involve the

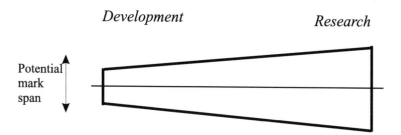

Development   *Research*

Potential mark span

Figure 11.1 The 'spectrum' of projects. Development projects should lead to a safe, passing mark for the student. Research projects are more risky. They are the ones that will attract first-class marks if they are done very well, but they can be difficult to write up if they do not go well.

implementation of a piece of software of a kind that has been implemented before (possibly many times), such as another database. There is little risk in this kind of project: as long as you do a reasonable job, you should be safe in reaching a passing mark. However, you are unlikely to achieve an exceptional mark. It is difficult to write up the implementation of a piece of software in such a way as to make it interesting or exciting. Development projects are good for students who are confident of their practical skills (for example, programming) but who are not aiming for an exceptional (that is, first-class) degree.

Research projects do involve some novelty. They are genuine research questions, that is to say something that has not been done before. That implies that the outcome is uncertain. The objective might be to prove some proposition but, by the very nature of the project, it is not known at the outset whether it will be possible to achieve that proof. Indeed, the proposition might turn out to be false. A proof of falsity is equally valid scientifically, but more difficult to write up positively. In other words, there is a greater risk attached to a research project. It might be that it turns out not to be possible (within the constraints of a student project) to prove or disprove the proposition. In that case, there is a danger that the project cannot be written up in an interesting style.

There is generally an implicit maximum mark available in a development project, and that is less than 100 per cent. That may seem unfair, but if put in the context of the whole range of projects it is quite fair. A student who takes on a more challenging (research) project and does well in it will deserve a higher mark than a student who undertook something rather more mundane and predictable. The important point is to realise the importance of choosing the right kind of project for you. By your final year you will be quite clear as to what kind of Computer Scientist you are – and also as to what kind of degree you can expect. If you are very committed to Computer Science, you are fairly certain you want a job in the IT industry or perhaps want to do research and you are hoping for a first-class degree, then you will want to do a research project. If you are glad that you are going to gain a Computer Science degree, but you think you may move on to some other kind of work and expect a 2(ii) degree, then you might be better thinking about a development-style project.

Of course, the point about Figure 11.1 is that it represents a scale. The discussion above has concentrated on the extremes of the spectrum, but there are points in between. A 'development' project might be based mainly on the production of a piece of software, but some research may be necessary to make it feasible to implement some component of that software. Such a project would be placed more to the right in the spectrum. Similarly, a 'research' project might also involve the implementation of some software, and in the event of the research not producing a clear result, the student might have to fall back on documenting that software as their main contribution in their project report. That project would be to the left of the research end of Figure 11.1. In defining a research project (in consultation with your project supervisor), it is often a good idea to have two versions of the project in mind. The first version is the one you are confident that you can achieve well within the time allowed. You know that if you complete that (sub-) project, and write it up well, then you should get a passing mark. The second project is an extension of that one. You would set out to achieve this more ambitious goal once you have completed the easier part of the project. If you achieve the goal, all is well and you can expect a high mark – but if you fall short, then it is not a disaster.

Once you have chosen a project and started on it, then good time management becomes vital (once more), and you must set out a timetable for the whole project. It is often difficult to estimate how much time different tasks need. For example, in a software project, it is notoriously difficult to estimate how much time it will take to write – and debug – a program. Nevertheless, it is important to make out even a rough timetable. It will allow you to chart your progress and to notice if you are falling behind. Whatever amount of time is allocated to the project (one semester, two, or even more) at the start, it will seem like a vast span ahead of you, with plenty of time to achieve all you have to. However, you will be surprised at how quickly that time passes. At the half way stage, most students look back and regret the amount of time that they have wasted. A realistic timetable, which assumes that you will devote your full effort throughout the period, should help you to avoid this pitfall.

It is vital that you include sufficient time in your schedule for the writing of the report. Students usually choose to leave the writing to the end, and this is a perfectly acceptable way of working, as long as you make good notes as you go along. Leaving the writing until the end also means there is a danger that if the implementation part of the project falls behind schedule, the student will be tempted to try to make up the time writing up the project in a shorter time. This is always a bad idea. You must include in your timetable a realistic period of time for the writing, and treat it as sacrosanct: whatever adjustments you may make to your schedule as the project proceeds, do not reduce the writing-up time. Many project reports give the reader the feeling that the student has worked long and hard on the practical work, but has rushed the report, and hence the project does not get the mark that the student might have deserved.

The assessment of the project will probably depend to a large extent on the report. There may be other elements as well, though. You may be required to give a presentation based on your project or a demonstration of any artefact that you developed.

Project work involves skills and tasks that you may not have practised much in your other studies. You will probably have to spend time in the library. You may have to look up research papers and not just textbooks. Your project report will probably be the most substantial piece of writing you have ever done – the kind of prose writing you have probably not done in your taught modules.

The subject of doing projects could be the topic of a book in itself – indeed, Dawson (2000) and Hunt (2005) are such books, and you should find them useful when you are preparing to undertake a project.

11.7 GROUP WORK

Another aspect of education that is unusual is the emphasis on individual work. In the 'real world' most people work in teams, and it is only appropriate that university courses should give experience of team working. The way group work is organised and assessed will differ between departments and modules.

There will be different roles in any group. Sorting out the roles is an important starting point. Some people may be more involved with design, some with implementation and others with evaluation. Any real group will generally have a leader. Within a student team, one or more assertive members may emerge – or even be singled out – but they will not generally have the power and authority of a manager in industry. In other words, they cannot dismiss team members.

There remains a tension between the desire for students to do group work and the fact that the eventual outcome is a degree classified for each individual. How should the marks be allocated between a team? You will often find that the lecturer assigns most of the marks to individuals, but there can be a portion of the marks to be shared among the team members in order to reflect their contribution to the result. It is not unusual for the team members themselves to be allowed to agree this allocation.

Student teams can sometimes be difficult to work in. The most common cause for complaint is members being seen as not pulling their weight. Giving team members some of the marks to distribute allows them to reward and punish, but in turn it can lead to further disagreements.

Nevertheless, even when it is not entirely positive and enjoyable, group work can be a very valuable experience.

12 ASSESSMENT

12.1 MARKS AND STANDARDS

One important distinction between university and sixth form that you should be aware of as soon as possible is the different level of marks that you can expect to attain. Honours degrees are usually given in classes, and the usual mark boundaries for them are shown in Table 12.1. For the really exceptional students, first-class degrees are sometimes awarded *with distinction* (often referred to as 'starred firsts', a bit like a starred A at 'A' level). The criteria for the award of distinctions varies between universities and is not necessarily a simple percentage score that could be added to Table 12.1.

Class	Abbreviation	Mark range (%)
First	1	70–100
Upper second	2(i)	60–69
Lower second	2(ii)	50–59
Third	3	40–49

Table 12.1 Honours degree classes and the marks to which they correspond. In addition, many universities will award an ordinary (non-honours) degree to a student who attains a mark of over 35 per cent

You will immediately notice in Table 12.1 that you can pass a degree with a mark as low as 35 per cent, and that overall the marks expected are somewhat lower than you are probably expect to get for your sixth-form work. There are a number of reasons for this. For one thing, as a potential university student your relative performance at school and sixth form must be high; in university you will be working with others of an approximately equal standard and you cannot all be given top marks. Also, university courses tend to cover a lot more material. You cannot be expected to remember all that in depth, so these marks demonstrate more of a breadth of understanding.

It is important that you are aware of these marks. Do not be downhearted if numerically your marks seem to drop in university assessments. Compare them to the above table, not to what you achieved at sixth form!

You will generally undertake exams and other assessments throughout the duration of your degree programme. Whether the marks you get throughout the programme count towards your final degree mark will vary between institutions. There will generally be a requirement to pass your first-year exams and assessments in order to enter the second year. In some departments that will be the only requirement, but in others your first-year mark will count towards your final degree classification. This will probably be a relatively small, weighted component of the final mark, though. In that way, your first year means something but it will not blight your chances of getting a good degree if you do not do well. It may take you some time to settle in to university study and you may do much better after the first year, ending up with a good class of degree, despite a 'slow' start.

Exam results are ultimately decided by a Board of Examiners. This will consist of the lecturers from the department, who have set and marked the exams, but it will also include some *external* examiners. These are lecturers from other universities who are brought in to oversee the examinations process. This is the mechanism whereby standards are maintained. A university cannot attempt to push itself up the league tables by simply relaxing its standards and by awarding more first-class degrees, because the external examiners would not allow it. Any decisions are made by the board of examiners as a body, but in controversial cases, they will usually defer to the externals.

You should also be aware that the boundaries in Table 12.1 are not hard and fast. When deciding degree classifications, examiners will look very carefully at borderlines. They are unlikely to be hard on a student who, after three years of work, is just, say, 0.4 per cent below a borderline. As Computer Scientists, they know that fraction-rounding errors can be that large, and that it would be unfair to penalise a student for inaccuracies in the arithmetic. Besides, no one can mark exams to perfect precision.

Given the discretion that examiners have, there are two lessons to bear in mind. First, if there is any circumstance in your life which may have adversely affected your performance in your exams or assessments (such as illness or bereavement), then make sure that you use whatever mechanisms there are in place to make this known to your tutor – and hence to the examiners. Secondly, it is a good idea to be known to your lecturers. A student who is recognisable as being one who asks intelligent questions in lectures, is more likely to get the benefit of the doubt at an examiners' meeting, than one whom none of the lecturers can even recall having seen in their lectures.

12.2 EXAMS

There will inevitably be some exams taken under closed conditions as part of your course. Having done SATs, GCSEs and ASs and possibly even 'A' levels, you will not be new to examinations, but there are nevertheless aspects of doing exams at university that you should be prepared for.

Exams are a peculiar phenomenon. There is nothing quite like them in 'real life'. Where else would you be expected to provide information entirely out of your head, with no reference to texts or other sources, and against a clock? The main reason exams persist, though, as a means of assessment is that it is the only safe method whereby the examiner can be quite certain that what is in front of them is the work of the individual student. Exams are also easier to mark than most other forms of assessment, and that is particularly important in large classes.

There are variations on the traditional closed, 'brain dump' exam. For instance, there are varieties of *open book* exam, in which you are allowed access to textbooks. The details of how these are run and administered will vary between departments.

Ideally, if you have followed all the above advice, you will have followed the material of all your modules as they have proceeded. You should aim for a deep understanding of the material, not just a surface understanding that you hope will get you through the exam.

The simplest advice on taking an exam is:

Read the question.

You have probably heard this before and, as a veteran of many exams, you probably think it is obvious – yet even graduating final-year undergraduates still make the mistake of not reading questions properly. If you do not read the question, then you are not likely to answer it. There is little that is more frustrating for an examiner marking a script than to have an answer in front of them that does not address the question. They will wonder whether the student genuinely misread the question, or did they simply not know the answer to the question in front of them and therefore use the politician's ploy of answering a different question to which they *did* know the answer? Could the student have written a good answer to the question asked if they had read it properly? Whatever is the reason for the poor answer, if it does not address the question, it cannot be given the marks.

You must also look for the key directions in the question. If it says 'define', then give a definition. If it says 'illustrate with examples', then do just that, and don't mix them up. If it asks for a definition, then don't give an illustrated example. Look at the mark allocations too. If a question carries one mark, then filling a page with an answer is going to be a waste of time; you will get your one mark, but no more.

Do not be put off by wordy questions. A long question often has a short answer, and vice-versa. A long question is probably worded very carefully to lead you toward the answer that the examiner requires, while a short question may open a can of worms (for example, 'Machines cannot think. Discuss').

Ideally, an exam should be a combination of testing facts and understanding. In other words, there will be some pure memory tests ('Give the definition of') and some that require you to solve a novel problem in the exam room. Of course, the time is limited, so you are not going to be asked to tackle anything too complex. Often a question that looks unfamiliar can turn out to be something you have seen before dressed up in different terms. If you read the question carefully and think about what it really says, it may suddenly become familiar and easy.

Do not worry too much if you cannot answer all of a question. In particular, you may not be able or have the time to answer the last part of the question. Examiners often put that in as an 'added extra'. It is there so that the really able student has a chance to show what they can do and get a really high mark while other, mere mortals will not do so but nevertheless get a passing mark. Again, look at the mark allocation and realise that you can still get a good pass without the five marks (or whatever) allocated to that final part. At the same time, do not be put off: writing something in answer to that hard question is infinitely more likely to get a mark than writing nothing!

One good method of preparing for an exam is to think about its content. Put yourself in the place of the examiner. Setting exams that are fair and a good test of knowledge and understanding is not an easy task. Look at the curriculum for the module concerned and identify the topics within it. Then pick out the ones that look as if they might form the basis of an exam question. Finally, think of some questions that a student could reasonably be expected to do under exam conditions and that would meet the other requirements of being a fair test. Most useful at this point, though, is to consider the other factors that are important to examiners. They want a question that is easy to mark: that is, it should be easy to allocate full marks, to a good answer, there is scope for partial marks and it is clear what is a wrong answer, and this decision should be possible in the shortest time to speed up the tedious marking process.

That may sound cynical and you should be aware that universities take their examination processes very seriously. There will be quality control mechanisms, such as moderation of marking, second marking and/or published marking schemes whereby the accuracy and fairness of all marking is checked.

Another useful form of preparation is to look at past papers. These can give you several sources of information. For a start, exam papers can be very similar from year to year. It is worth picking out the patterns. Indeed, some lecturers will deliberately put questions that are identical in structure and only vary in the data in every paper. They are making a point: these are basic facts that students every year are expected to learn. Then there may be repetition because there are only so many questions that can be asked on a topic (as you may be aware if you have tried the above exercise of writing your own exam paper). The fact that a reduced set of questions is likely to be asked is useful information. Beware though, that to go further and to try to play the game of anticipating which questions will be included this year is rather more risky. Finally, past papers give you a chance to practise. You should try to do at least one paper under fully simulated conditions – all in one go in the allotted time and without reference to any books.

Departments vary as to their policy on releasing model solutions. While it is helpful for students to be able to compare their answers with those of the lecturer, some may be reluctant to publish solutions, particularly for exams which do not vary much from year to year. If a solution is available, you really should have a go at doing the paper yourself before you look at the answers. It is easy to convince yourself that you knew the answer when it is in front of you. When you do compare answers, some of them will be obviously right or wrong (for example, those that are mathematical) but there may be others where your answer differs from the published one, but is not necessarily wrong. Remember they are *model* answers and there is not always a right or wrong response. Comparing two such answers is a very good test of whether you understood the material – and the question.

If, despite all you have read here and all your hard work, you should have the misfortune to fail one or more exams, all is not lost. Often if you fail an exam you will be given a chance to resit it. If you fail exams during your first year, you may be allowed to resit them at some time over the summer, and if you pass the resit, you will be allowed into the second year.

There are a few points to note about resits. One is that they are best avoided. If you do the work, you should pass first time and avoid the stress and inconvenience of having to take more exams. Secondly, you should be aware that resits are sometimes treated as more of a privilege than a right. In other words, the chance to take a resit may not automatically follow the failing of an exam. Thirdly, the way the marks for resits are calculated is likely to be done in such a way that you will not get a better mark having taken a resit than other students did taking it another time. Finally, another disincentive for resits is that you may be charged a fee for taking them. You may also have to pay for accommodation in the university for the duration of your resits.

12.3   *VIVA VOCE* EXAMINATIONS

In addition to written examinations, most departments retain the right to require individuals to undergo an oral examination (also known as a *viva voce* exam, or simply 'viva'). Usually, it will be the external examiners who will decide whether and whom to examine in this way. They will often use this as a mechanism to help them to decide which side of a borderline to put a candidate. Should you be requested to attend a viva, therefore, you can guess that your written exam and assessment results have put you near a borderline – but you will not know which borderline!

In principle, the examiners can ask you about any topic in Computer Science. In practice, they are most likely to ask you about your project (if you have done one). This is a chance for interesting discussion around a topic which you should know well.

There is not much you can do in preparation for such an open examination. You should probably read your project report one more time and you might like to think about the kinds of questions you are likely to get on it. Are there any omissions you are aware of or potential extensions? If so, it is to be hoped you listed these in a Further Work section in the report, but the examiners might take the chance to probe you further on exactly how you would envisage pursuing these new leads. You might also try to criticise your own work (again, it is even better if you have already done this in your report). Are there things that you would do differently if you were to start the same project again, knowing what you know now?

While trying to anticipate the criticisms that the examiners may try to make of your work, do not forget what is good about it. Remind yourself of what you achieved and take the chance to make sure that the examiners are aware of that too. The examiners will not be trying to catch you out and there will not be trick questions. What they will be looking for is a stimulating discussion that will give them the opportunity to slide you above whatever borderline you are on.

It is easy to say that you should try not to be nervous about a viva, and more difficult to achieve this. Prepare, by thinking about your project. Do not think of the examiners as hostile; think of them as being on your side. Be honest, but be prepared to 'sell' yourself as best you can.

12.4 COURSEWORK

While exams are artificial, many modules will have an element of assessment done under open conditions, which is rather more like the real world. In Computer Science, in particular, programming skills can only be truly assessed by the student writing programs. The way these are administered will vary between departments and modules. It may be that you have quite a short time to work on such an assessment, or it may take place over an extended period of a week or two. This is a real test of time management. There is a temptation to concentrate all your time on the assessment – the piece of work that actually counts towards your degree. You must not go too far down that road, though. If you stop everything else – such as going to lectures – you will fall behind in your other modules and they will suffer.

Programming assessments can be particularly beguiling. You become so engrossed in the problem that you get distracted from everything other than getting the program to work. You should have a breakdown of the marks for the assessment, so look at it carefully and work to it. It is often the case that getting the program to work carries a relatively small proportion of the marks. Documentation may be very much more important. It is rarely the case that another 5 per cent of time spent improving a program will get you 5 per cent more marks. A sign of a mature student is one who has learned when to stop writing and 'improving' program code and to spend his or her time on more productive work. It usually takes several – largely wasted – all-night programming sessions before this lesson comes home.

Increasingly, a problem that universities have with coursework assessments is students copying all or parts of their submissions. Passing off the work of another as your own is called plagiarism. Don't do it.

Plagiarism takes different forms. Increasingly, people find pieces of work on the Web which address a question similar to the one they have been set. There are even websites set up explicitly for that purpose. They may copy from other students – with or without the collaboration of the other student. There are several reasons why you should not give in to the temptation to plagiarise, including:

It is cheating
When you graduate, you want to know that achieved got the class of degree you deserved.

You will probably be caught
Most lecturers can easily spot copied material. There are invariably telltale signals such as sudden changes in style, segments that

fit together badly or identical errors in different people's answers. Also, just as the technology can be used to create plagiarism, it can be used to detect it. A few minutes' work on Google can often reveal the original source of plagiarised material. Beyond that, though, many departments run student work through plagiarism detection tools as a matter of course.

The punishments are harsh

Universities have a range of punishments they can apply. You may be removed from the course and you may have the award of any degree withheld. Imagine phoning home with that news.

Beware that you should not take part in plagiarism at all. If you did the work yourself but then allowed a colleague to copy from you, you are likely to be treated as being as guilty as they are.

If you have read from the beginning of this book to here, then you should be in a good position to apply for, be accepted and start on a degree in Computer Science. Enjoy the ride! It can be a great time. You should make the most of it and be well set up for a rewarding and enjoyable career.

GLOSSARY

Algorithm: A set of instructions that can be followed mechanically to solve a problem. An algorithm is similar to a program but is expressed in a way that is independent of any particular implementation. That is to say, an algorithm does not rely on the features of particular programming languages or of any particular computer.

Binary: The base 2 number system. It consists of two digits: 0 and 1, referred to as *bits*. It is the basis of **digital** computer systems because they can easily represent these simple, discrete values.

Brute force: A simple – but inefficient – approach to problem solving. Suppose, for instance, you wanted to find the shortest route between two towns and knew the distance between a number of alternative intermediate towns. You could simply trace all the possible routes and then find which is shortest. This would be a brute-force solution. The problem is that if there are a lot of interconnected intermediate towns, then the total number of possible routes can be vast. There could be so many that the brute-force approach might be impractical, and a more refined method might be sought. On the other hand, the increasing power of computers means that brute-force approaches that were not practical with previous generations of computers are now sometimes feasible.

Code: Short for 'program code', and a term used to refer to computer programs. 'To code' can be a synonym for 'to program'.

Compiler: A program that processes other programs. They will be written in a 'high-level' language that is readable by programmers, but must be converted into a low-level language (essentially a series of binary 1s and 0s, known as machine code) that can control the computer processor. The whole of the input program is converted at one go by the compiler. (See also **interpreter.**)

Computer engineering: This term is generally used to refer to the design and building of computer **hardware.**

Determinism: See **non-determinism.**

Digital: Technology based on the **binary** number system. The ability to transform almost any information into a digital form – and the availability of technology to handle that form – explains much of the power and usefulness of modern technology.

Hardware: The physical components that make up a computer. 'Hard' implies that these components are fixed and cannot be changed, whereas the **software** that runs on the computer can be changed to make it do different things.

ICT: Information and Communication Technology, or Information and Communication Technologies. These terms refer to the fact that computer technology and the technology for connecting computers together (networking) have developed in an inter-dependent manner. (See also **IT.**)

Information technology: See **IT.**

Interpreter: A program that runs other programs. The other program (**source code**) has been written in a high-level programming language that is understandable by people but cannot run directly on the computer **hardware.** Instead, the interpreter executes the instructions of the source code. This is similar to a **compiler,** except that it is undertaken interactively; the interpreter runs its input program line-by-line. Basic is an example of a language that is usually interpreted. However, Basic compilers also exist. The important difference is that a compiled program will execute more quickly.

IT: Information Technology. This is used as a general term to refer to computers and related technology, reflecting the fact that the technology is often broader than just computers. (See also **ICT.**)

   IT is also used as the name of some degree programmes, in which case it generally refers to something different from

Computer Science. For a fuller explanation, see Section 8.1: Computer Science, Studies, IT . . .?

Moore's Law: The observation that the power of computer chips doubles approximately every eighteen months. This 'law' is named after Gordon Moore, who first made the observation in 1965 (Moore 1965). So far, the observation has continued to hold true.

Non-determinism: The term *non-determinism* (and *determinism*) is used in two ways. The first instance is sometimes referred to as *oracular non-determinism*, referring to the idea that the 'oracle' can be consulted and it will guess the right answer. The other use of non-determinism is more general, more akin to randomness, whereby it is not possible to predict which option will be chosen at any decision point. By either definition, all real computers are *deterministic*. That is to say that at any point, knowing the state of the computer and its current input, it is possible to say precisely what its next step will be.

Object code: The lowest-level form of a computer program, essentially a set of binary 0s and 1s that can control the computer processor. (See also **compiler** and **interpreter**.)

Object oriented: A particular approach to programming, also applied to design. Programs are constructed from objects, which embody information and the operations that can be applied to that information.

Pixel: A digital picture is made up of a set of dots, known as pixels or 'picture elements'. Each pixel is represented by one or more numbers. The density of pixels dictates the resolution or sharpness of the picture, which is why digital camera advertisements boast about the number of megapixels per (square) inch.

Program: A set of instructions for a computer to make it perform a particular set of actions.

Real time: Events in the world take place in 'real' time. Sometimes computer programs have to keep up with those events. For example, the autopilot in an aircraft must take avoiding action before the aircraft hits an obstacle such as a mountain. If the software takes too long to calculate the aircraft's position, then there could be a disaster.

Software: A collective name for computer **programs**, used to distinguish them from **hardware**. 'Soft' implies that programs are relatively malleable and can be changed with relative ease.

Software engineering: The application of engineering techniques to the construction of computer programs.

Source code: A computer program expressed in a 'high-level' language, one which can be read and understood by people. In order to run on a computer, it must be translated into **object code** by a **compiler** or **interpreter**.

Syntax: The grammar of a language which defines legal combinations of symbols in that language. In Computer Science, syntax often refers to the grammars of (artificial) programming languages, but within the sub-discipline of Natural Language Processing, it may also refer to spoken languages such as English or French.

Virus: Strictly speaking a virus is one particular kind of program that has been written specifically to spread and damage computer systems. There are many different kinds of such software, including viruses, *Trojan horses* and *worms*, but the term *virus* is often used as a collective term to cover all types.

Wysiwyg: 'What you see is what you get', pronounced 'wizzy-wig'. Originally this term applied to the display of documents on a computer screen, the appearance of which was identical to how the document would look when printed on paper. The term is often generalised to any situation in which there is a direct link between an object and its representation.

REFERENCES

Abelson, H., Sussman, G. J. and Sussman, J. (1985) *Structure and Interpretation of Computer Programs*, Cambridge, MA: MIT Press.

Ayres, R. (1999) *The Essence of Professional Issues in Computing*, London: Prentice Hall.

Baeza-Yates, R. and Ribeiro-Neto, B. (1999) *Modern Information Retrieval*, Reading, MA: Addison-Wesley.

Becker, L. and Price, D. (2003) *How to Manage Your Science and Technology Degree*, Basingstoke: Palgrave Macmillan.

Berger, A. (2001) *Embedded Systems Design: A Step-by-Step Guide*, New York: Osborne McGraw-Hill.

Bernstein, P. A. (1996) 'Middleware: a model for distributed system services', *Communications of the ACM* 39(2): 86–98.

Birtwistle, G. M. (1979) *Discrete Event Modelling on Simula*, Basingstoke: Palgrave Macmillan.

Booch, G. (1994) *Object-Oriented Analysis and Design*, New York: Addison-Wesley.

Britton, C. and Bye, P. (2004) *IT Architectures and Middleware*, Boston: Addison-Wesley.

Burns, A. and Wellings, A. J. (2001) *Real-Time Systems and Programming Languages*, Reading, MA: Addison-Wesley.

Callan, R. (1998) *The Essence of Neural Networks*, London: Prentice Hall.

Carter, J. (2003) *Database Design and Programming with Access SQL, Visual Basic and ASP*, London: McGraw-Hill.

Cawsey, A. (1997) *The Essence of Artificial Intelligence*, London: Prentice Hall.

Chapman, N. and Chapman, J. (2004) *Digital Multimedia*, Reading: John Wiley.

Clements, A. (2000) *The Principles of Computer Hardware*, Oxford: Oxford University Press.

Cooley, P. (2000) *The Essence of Computer Graphics*, London: Prentice Hall.

Coulouris, G., Dollimore, J. and Kindberg, T. (2005) *Distributed Systems: Concepts and Design*, Reading, MA: Addison-Wesley.

Crichlow, J. M. (2000) *The Essence of Distributed Systems*, Harlow: Pearson Education.

Currie, E. (1999) *The Essence of Z*, London: Prentice Hall.

Date, C. J. (2003) *An Introduction to Database Systems*, Reading, MA: Addison-Wesley.

Dawson, C. W. (2000) *Computing Projects: A Student's Guide*, Harlow: Prentice Hall.

Dean, N. (1996) *The Essence of Discrete Mathematics*, London: Prentice Hall.

Deitel, H. M. and Deitel, P. (2000) *E-Business and E-Commerce: How to Program*, London: Prentice Hall.

Dix, A. (2002) 'The ultimate interface and the sums of life?' *Interfaces* 50 (Spring): 16.

Dix, A., Finlay, J., Abowd, G. and Beale, R. (2003) *Human-Computer Interaction*, London: Prentice Hall.

Edwards, A. D. N. (1991) *Speech Synthesis: Technology for Disabled People*, London: Paul Chapman.

Edwards, A. D. N. (ed.) (1995) *Extra-Ordinary Human-Computer Interaction: Interfaces for Users with Disabilities*, Cambridge Series on Human–Computer Interaction, New York: Cambridge University Press.

Fabris, P. (1998) 'Advanced navigation', *CIO Magazine* (http://www.cio.com/archive/051598_mining.html).

Faulkner, X. (1998) *The Essence of Human-Computer Interaction*, London: Prentice Hall.

Foley, J. D., van Dam, A., Feiner, S. K., Hughes, J. F. and Phillips, R. L. (1994) *Introduction to Computer Graphics*, Reading, MA: Addison-Wesley.

Gaver, W. W. (1997) 'Auditory interfaces', in Helander, M. G., Landauer, T. K. and Prabhu, P. (eds) *Handbook of Human-Computer Interaction*, Amsterdam: Elsevier Science.

Giodano, F. R., Weir, M. D. and Fox, W. P. (1997) *A First Course in Mathematical Modeling*, Pacific Grove, CA: Brooks/Cole.

Griffiths, G. (1998) *The Essence of Structured Systems Analysis Techniques*, London: Prentice Hall.

Horowitz, E. (1987) *Fundamentals of Programming Languages*, Maryland: Computer Science Press.

Hunt, A. (2005) *Your Research Project: How to Manage It*, Oxford: Routledge.

Hunter, R. (1999) *The Essence of Compilers*, London: Prentice Hall.

Inmon, W. H. (1996) *Building the Data Warehouse*, New York: Wiley.

Johnson, D. G. and Nissenbaum, H. (2004) *Computers, Ethics, and Social Values*, Englewood Cliffs, NJ: Prentice Hall.

Kelly, J. (1996) *The Essence of Logic*, London: Prentice Hall.

Kendall, J. E. and Kendall, K. E. (2002) *Systems Analysis and Design*, Upper Saddle River, NJ: Prentice Hall.

Knuckles, C. D. and Yuen, D. S. (2005) *Web Applications: Concepts and Real World Design*, Hoboken: Wiley.

Knuth, D. E. (1973) *The Art of Computer Programming*, Reading, MA: Addison-Wesley.

Lunn, C. (1996) *The Essence of Analog Electronics*, London: Prentice Hall.

McGettrick, A., Arnott, J., Budgen, D., Capon, P., Davies, G., Hodson, P. J., Hull, E., Lovegrove, G., Mander, K. C., McGrath, P., Norman, A., Oldfield, S. J., Rapley, A., Rayward-Smith, V. J., Simpson, D., Sloman, A., Stowell, F. A. and Willis, N. (2000) *Computing*, Gloucester: Quality Assurance Agency for Higher Education Subject Benchmark Statement.

Moore, G. E. (1965) 'Cramming more components onto integrated circuits', *Electronics* 38 (8).

Nielson, H. R. and Nielson, F. (1999) *Semantics with Applications: A Formal Introduction*, Chichester: John Wiley.

Northedge, A., Thomas, J., Lane, A. and Peasgood, A. (1997) *The Sciences Good Study Guide*, Milton Keynes: Open University Press.

Paulson, L. C. (1996) *ML for the Working Programmer*, Cambridge: Cambridge University Press.

Pitt, I. and Edwards, A. (2002) *Design of Speech-based Devices: A Practical Guide*, London: Springer.

Ransom, R. (1987) *Text and Document Processing in Science and Technology*, Wilmslow: Sigma Press.

Ratzan, L. (2004) *Understanding Information Systems: What They Do and Why We Need Them*, Chicago: ALA Editions.

Rolland, F. D. (1997) *The Essence of Databases*, London: Prentice Hall.

Russell, D., Gangemi, G. T. and Gangemi, G. T., Sr (1991) *Computer Security Basics*, Sebastopol, CA: O'Reilly.

Russell, S. J. and Norvig, P. (2003) *Artificial Intelligence: A Modern Approach*, Upper Saddle River, NJ: Prentice Hall.

Schalkoff, R. (1989) *Digital Image Processing and Computer Vision*, New York: John Wiley.

Sebesta, R. W. (1999) *Concepts of Programming Languages*, Reading, MA: Addison-Wesley.

Sedgewick, R. (1988) *Algorithms*, Reading, MA: Addison-Wesley.

Shneiderman, B. (1998) *Designing the User Interface: Strategies for Effective Human-Computer Interaction*, New York: Addison-Wesley.

Sloman, M. and Kramer, J. (1987) *Distributed Systems and Computer Networks*, New York: Prentice Hall.

Smith, S. L. and Mosier, J. N. (1984) *Design Guidelines for User-System Interface Software*, Hanscom Airforce Base, MA: USAF Electronics Division ESD-TR-84-190 (http://www.info.fundp.ac.be/httpdocs/guidelines/smith_mosier/SM.html).

Sommerville, I. (2004) *Software Engineering*, New York: Addison-Wesley.

Stallings, W. (2003) *Computer Organization and Architecture*, Upper Saddle River, NJ: Prentice Hall.

Tanenbaum, A., Day, W. and Waller, S. (2002) *Computer Networks*, Englewood Cliffs: Prentice Hall.

Tanenbaum, A. S. (2001) *Modern Operating Systems*, Englewood Cliffs: Prentice Hall.

Thompson, S. (1996) *Haskell: The Craft of Functional Programming*, Reading, MA: Addison-Wesley.

Truss, L. (2003) *Eats, Shoots and Leaves: The Zero Tolerance Approach to Punctuation*, Profile Books.

Turing, A. (1936) 'On computable numbers with an application to the Entscheidungsproblem', *Proceedings of the London Mathematical Society* 42: 230–65.

Wellings, A. (2004) *Concurrent and Real-Time Programming in Java*, Hoboken, NJ: Wiley.

Winston, P. and Horn, B. K. P. (eds) (1989) *Lisp*, Reading, MA: Addison-Wesley.

Ziemer, R. E. and Tranter, W. H. (1995) *Principles of Communications: Systems, Modulation and Noise*, Boston: Houghton Mifflin.

INDEX